STRAIGHT SHOOTER

A game-changing new approach to basketball shooting

Bob J. Fisher

Dedicated to the young man or woman
looking for the answer to better shooting.

"There may not be a more learned student of the jump shot than Bob. In his latest book, perhaps the greatest free throw shooter of all time begins exploring the traditional shooting teachings and snowballs that information with deeper data and modern advancements. Bob's work crescendoes perfectly into his nearly-perfected teachings and philosophy."

Jake Fischer
Associate Producer
Sports Illustrated

"Bob does a great job of simplifying the intricate task of shooting a basketball. He has taken years of research and trial and error to develop a simplistic style that will help anyone become a better shooter."

Brad Underwood
Head Men's Basketball Coach
University of Illinois

"Kids today need to understand basic fundamental skills of shooting. This book will further their success."

Bill Cartwright
Retired NBA player (16 seasons)
Five-time NBA Champion with the Chicago Bulls
NBA and international head coach

"Bob Fisher has studied the art of shooting more than anyone I know. He is also one of the most accurate shooters I have ever seen. Who wouldn't want to read his book and learn his philosophy on shooting?"

Jackie Stiles
Scored NCAA record 3,393 points
WNBA Rookie of the Year and WNBA All-Star
Assistant Coach—Missouri State Lady Bears

"Coach Fisher does a good job of breaking down the overall shooting technique. You can take the things he teaches and see immediate improvement in your shooting."

Bill Self
Head Men's Basketball Coach
University of Kansas

"*Straight Shooter* will do just that—straighten out your shooting. By taking an easy-to-understand, physics-based approach to basketball, Fisher will improve many facets of your game. After all, it all starts in the mind—but it will end up in the basket."

Philip Reed
Co-author of *Free Throw, 7 Steps to Success at the Free Throw Line*

"Bob Fisher is a perfect example of how world-class talent can be built with creativity, hard work, and a relentless spirit."

Daniel Coyle
author of *The Talent Code*

"Your book is a WINNER! You have a talent nobody has ever had, but every basketball shooter wants."

Jeff Liles
19 time Guinness World Record free throw shooter

"Shooting is such a weird thing because coaches often just re-hash what they have heard from other coaches or have always been taught. Everything in the game is constantly changing and enhancing itself, except shooting. Why is that? Bob's centerline method is simple and something I have never heard anyone teach. It's refreshing to see someone dive into it instead of saying square your shoulders, flick your wrist, keep your elbow in, etc.—all the dumb coach talk that goes along with shooting a basketball that has no impact on someone's shot."

Pat Schulte
Men's Basketball Video Coordinator
University of Illinois

All rights reserved. No part of this publication may be reproduced, distributed, or transmitted in any form or by any means, including photocopying, recording, or other electronic or mechanical methods, without the prior written permission of the publisher, except in the case of brief quotations embodied in critical reviews and certain other noncommercial uses permitted by copyright law.

Cover illustration by Gunnar Assmy/Shutterstock.com

Grateful acknowledgment for permission to reproduce illustrations is made to the following:
Aatish Bhatia, page 129; Larry Silverberg, page 150.
Unless otherwise noted in the text, all photos are by Connie Fisher.

Published and distributed by: Telemachus Press, LLC
7652 Sawmill Road
Suite 304
Dublin, Ohio 43016

ISBN: 978-1-945330-82-7 eBook
ISBN: 978-1-945330-82-7 Paperback
ISBN: 978-1-948046-13-8 Hardback

Library of Congress Control Number: 2018954864

http://www.telemachuspress.com

Visit the author website:
www.secretsofshooting.com

Categories:
SPORTS & RECREATION / Basketball

2018.12.07

Bob Fisher World Records* and Personal Bests

Most Basketball Free Throws Made:

In one hour (2371)
In 30 seconds (33)
In one minute (52)
In two minutes (92)
In 10 minutes (448)
In one minute blindfolded (22)
In two minutes blindfolded (37)
In one minute underhanded (28)
In one minute while alternating hands (45)
In two minutes while alternating hands (100)
In one minute by a pair–with Jeff Liles (46)
In one minute with a coed pair–with Brandi Jo Roepke (44)
In one minute by a pair with two balls–with Jeff Liles (33)
In one minute while standing on one leg (49)

Personal Bests:

Made 10 free throws in 5.04 seconds
Made 41 free throws in 30 seconds
Made 58 free throws in one minute
Made 21 free throws in 30 seconds while blindfolded
Made 29 free throws in one minute while blindfolded
Made 47 free throws in one minute while alternating hands

* As recognized by *Guinness World Records*™; a Trademark of Guinness World Records Limited

Table of Contents

Introduction — i

Part 1 — 1
Chapter 1: The Problem with Mechanics — 3
Chapter 2: Uncharted Waters — 13
Chapter 3: The Secret of the Release — 28

Part 2 Special Section by Larry Silverberg: — 61
Physics of the Free Throw

Part 3 — 87
Chapter 4: Objectives — 89
Chapter 5: Be a Straight-shooter — 94
Chapter 6: The ANSWER — 110
Chapter 7: Funneling — 121
Chapter 8: Long and Short — 128
Chapter 9: Ways to Miss — 134
Chapter 10: Controlling Momentum — 139
Chapter 11: Sidespin — 144
Chapter 12: Bank — 148
Chapter 13: Free Throws — 153
Chapter 14: Lateral Movement — 160
Chapter 15: Downtown — 162
Chapter 16: Dr. Tom — 166
Chapter 17: Tall vs. Short — 173
Chapter 18: High or Low — 177
Chapter 19: Exercises — 182
Chapter 20: Closing Comments — 185

Acknowledgments — 197

About the Author — 199

STRAIGHT SHOOTER

A game-changing new approach to basketball shooting

Introduction

"We see what we look for and we look for what we know."
                                                                Goethe

Be warned; in this book, I deviate from the norm. But there is a reason for the deviation. I often start shooting clinics by saying, "Don't believe everything you have heard about how to shoot a basketball. Everyone has an opinion on shooting and there is a lot of misinformation floating around. You shouldn't believe anything I tell you here today either unless I can back it up with physics or some scientific study."

Although I hold 14 Guinness World Records™ in free throw shooting, my claim to fame is I can knock down shots faster than anyone. I have made ten free throws in 5.04 seconds; 41 in 30 seconds; and 58 in one minute. The reason I can shoot accurately at a rapid pace is that I utilize a different approach. The traditional approach to shooting emphasizes fundamental mechanics. What I will introduce to you in the following pages is a new physics-based approach which is far superior and will help you improve quicker.

(Photo courtesy of Steve Ridgeway of NBC)
On The Tonight Show with Jay Leno
and guest NBA Hall-of-Famer Charles Barkley.

Being a world record holder has provided some unique opportunities. I shot against Charles Barkley on *The Tonight Show with Jay Leno*; was featured on *Inside Edition*; traveled to China to shoot

i

on the *Zhang da Zhong Ye Show*; taught shooting in the documentary *Free Throw* at Compton High School in Southern California, and demonstrated shooting at the New York World Science Festival. John Branch, a sports writer for the *New York Times*, traveled to Kansas to watch me break six Guinness World Records™ in one hour. Branch's article made the front-page.

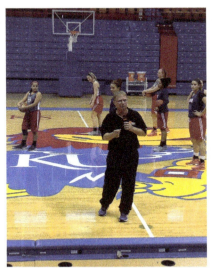

Giving a shooting presentation at the KU Coaches Clinic, fall 2016.

Good Advice

Thirty years ago, I heard the advice, "Become an expert in something." I instantly knew I wanted to become an expert in shooting a basketball. However, I was 49 years old before I discovered the information which would lead me to develop a physics-based approach.

The catalyst which started my journey to enlightenment was John Fontanella's book *The Physics of Basketball*. It opened the door to a new and different perspective for me. Fontanella was a shooter in college, a guard at Westminster College in New Wilmington, Pa. He set the school record for points in a game (51) and still owns three of the top-five scoring games in school history. He also led the nation one year in free throw percentage by shooting an impressive 92%.

Different Releases

In an interview with National Public Radio, Fontanella said the most important aspect of shooting was the release and which fingers leave the ball last.[1] For him, that was the middle and ring fingers.

This information was new to me. For 49 years, I had been shooting with my index finger leaving the ball last. I immediately hit the gym to check out Fontanella's method. That began the search which resulted in this book. While experimenting with a vast array of shooting styles over the past ten years, I have developed and honed a new physics-based methodology to shooting a basketball. This approach will help *you* make shots.

These ideas will open the door to possibilities you never knew existed. You will improve faster and shoot better than you ever thought possible.

Physics-based Concepts

In this book, I will share this new approach and also teach you some physics-based concepts that will enhance your game. Great players learn these concepts intuitively, without an awareness of their specifics. Knowing what you are doing and why it works will enable you to utilize these methods to full advantage and gain you a considerable advantage over your opponents.

Larry M. Silverberg, Professor of Mechanical and Aerospace Engineering at North Carolina State University, is one of the world's leading experts on the physics of basketball. He has conducted a large number of numerical studies of the bank shot and the free throw.

Silverberg's positive reaction to this book resulted in the inclusion of the underlying physics of basketball shooting in a special chapter titled; The Physics of the Free Throw. For the reader's enjoyment, his chapter approaches shooting from an academic standpoint, explaining the background of physics and strengthening the themes throughout this book.

Game Shooting

A majority of the time, field-goal percentage is the determining factor in whether teams win or lose. The information presented here will help you outshoot your competition. This book is the first outlining of the physics-based approach that I have developed over the last ten years. These ideas will work for everyone because they are grounded in science, which allows you to use your natural biomechanics.

Immediate Results

By learning this material, you will be able to fast-track your shooting prowess. Techniques which take years for the pros to learn are broken down into segments you can quickly understand and apply. You can elevate your shooting to new heights. Plus, this physics-based approach will work for you *today*.

The truth is you can shoot numerous ways effectively. Some methods may come more natural to you than others. But they will all work. Ultimately, what techniques you choose to use is entirely your decision.

Using a physics-based approach will accelerate your development. This method is new and different from the standard strategy of trying to correct mechanics. It transforms shooting into the learning experience it should be.

Part 1

Chapter 1
The Problem with Mechanics

"It's not what we know that hurts us. It's what we know for sure that just isn't so."
Will Rogers

"The one thing we do know is the traditional approach and nothing else, of simply trying to correct mechanics and go in the gym and shoot a lot of free throws, has not worked." [1]

Stan Van Gundy, head coach of the Detroit Pistons, made this remark following the 2015-2016 season, about Andre Drummond's problems at the free-throw line. Drummond, a center for the Pistons, is an NBA All-Star.

Mechanical Breakdown

In 2015, Andre Drummond shot 38.9% from the free-throw line. Opposing teams were implementing the "Hack-A-Shaq" strategy and Drummond found himself on the bench at the end of close games. To keep the league's leading rebounder in the game, the Detroit Pistons brought in Dave Hopla, an acclaimed shooting coach.[2] Hopla is someone who can walk the walk because he is one of the best shooters in the world. Hopla has made over 1,200 free throws in a row and regularly makes 595 out of 600 free throws *while he is talking* at clinics.

Hopla has been an NBA shooting coach for years and has helped numerous players with their mechanics. He pays attention to technical matters; like how a player gets his legs into a shot, foot placement, elbow position, release point and follow through.

The next season Drummond's free-throw percentage dropped to 35.5%. On Jan. 20, 2016, in a game against the Houston Rockets, he set a record for missed free throws when he misfired on 23 of 36 attempts.

Oversell and Underdeliver

"I will completely change his numbers in less than one week and you won't recognize him." [3]

Ed Palubinskas made this claim in an email to the Orlando Sentinel newspaper in May 2010. His comment referred to the free throw shooting of Dwight Howard, the Orlando Magic's center. Palubinskas received the opportunity to back up his boast a year later. After the 2010-11 season, Howard announced he would hire a shooting coach. He interviewed several potential candidates before he hired Palubinskas. At this point, Howard was a career 59.8% free-throw shooter.[4]

Palubinskas bills himself as the "Shooting Surgeon General" and the "Free-throw Master." He played for Australia in the 1972 and 1976 Summer Olympics and is the career free-throw percentage leader at Louisiana State University. He worked with Shaquille O'Neal of the Los Angeles Lakers during the 2000-2001 season. (After the season, he and the Lakers parted ways when they rejected his request for a yearly salary and a championship ring.)[5]

Palubinskas started working with Howard in August. That fall, the NBA players went on strike. It delayed the start of the 2011-2012 regular season from Nov. 1 to Dec. 25 and reduced it from 82 to 66 games. When play resumed, Howard made 28 out of 67 free throws to start the season. Howard finished the season with a 49% free-throw percentage.

Worse Instead of Better

What happened? What's going on here? Hopla and Palubinskas are both acclaimed shooting coaches who know the mechanics of shooting. They are great shooters themselves, training two of the most athletic players in the league. Both players were receptive to instruction and there was tremendous room for improvement. How could their free-throw shooting get worse?

A Different Outcome

From these examples, one might assume shooting coaches aren't all they are cracked up to be. But consider the case of Ted St. Martin and the

Jacksonville University (Florida) Dolphins. St. Martin holds the world record for consecutive free throws made (5,221) and teaches shooting to kids at clinics.[6]

Late in the 1990-91 season, Jacksonville University coach Rich Haddad brought in St. Martin to work with the Dolphins, who were the worst free-throw-shooting team in the Sun Belt Conference. They were shooting under 50% for the season. St. Martin spent one practice with the players, working with each one individually for five to 10 minutes. By the end of the day, each player finished above 70%. The squad lifted its team percentage from below 50% to 70% by season's end.

A Question Arises

What did Ted St. Martin teach which resulted in such a drastic improvement—more than 20%—in just one practice session?

The answer will help you dramatically improve your shooting and also help you avoid "choking" when shooting a clutch free throw.

The Mechanics Approach

In his book *Basketball Shooting*, Hopla provides detailed instruction on shooting technique. "Once the shooting arm is in the L-shaped position, you should make sure that the elbow is directly under the ball and above the shooting foot. This L-shaped position is key." When describing the follow-through, Hopla advises, "The shooting arm is fully extended with the elbow above the eyebrow and slightly in front of the body. It should stay in the shot line: the toe, knee, hip, shoulder, elbow, wrist, and hand all maintaining alignment. The shooting hand should be pointed toward the target and in the hoop with fingers pointed down at about a 45-degree angle toward the floor." [7]

Palubinskas teaches the index finger should be in the center of the ball and the shot pocket should be right in front of the eyes and two to three inches in front of the forehead. According to Palubinskas, when you shoot your fingers should be spread wide with the hand square to the rim. On your follow-through, your fingers should be in the same position as when you do a fingertip pushup.[8]

There is much more to the teachings of both these men, but this gives you a flavor of how they teach. Both are specific and detail-oriented. They are masters at breaking down and explaining their particular shooting styles. Palubinskas is a strong proponent of the index finger in the center of the ball and believes it to be the best way to shoot. Hopla favors placing the index and middle fingers in the center of the ball.

One of a Kind

St. Martin never played in the NBA, nor did he play college ball.[9] At 5 feet 7 inches tall, he wasn't even the star of his high school team in Naches, Washington. Until the age of 35, St. Martin was a dairy farmer. One day he nailed a hoop to the side of the dairy barn and started shooting. After a while, St. Martin was able to hit a couple of hundred free throws in a row. One evening he proceeded to amaze family and friends by sinking 514 consecutive shots.

At the time, the record for most consecutive free throws was 144. In short order, St. Martin broke it with 200. He then proceeded to break it again and again: 245, 305, 315, 386, 927, 1,238, 1,704, 2,036 and ultimately 5,221 in 1996.

St. Martin was sponsored by AMF Voit, Asics, Adidas, and Coors. For over a decade, he traveled the country putting on demonstrations and taking on all challengers. St. Martin performed during the Final Four weekend, at NBA games, malls, charity events, conventions, carnivals and grand openings. After he retired to Jacksonville, Florida, St. Martin continued to give shooting clinics and teach shooting.

A Different Style of Teaching

What did Ted St. Martin do with the Jacksonville University players? How did he help them achieve such dramatic improvement?

It is imperative to know what he didn't do. St. Martin had an unorthodox shot. He brings the ball to his chest and releases a two-handed set shot starting at his chin. His left guide hand opens up and his fingertips flash outward. With hands on either side of the basketball, his hands don't wave the ball to the basket the way a fingertip release does. Instead, he shows the open palms of his hands at the moment of release.

Because of St. Martin's different-looking shot, he didn't try to teach the Jacksonville University players his technique. Nor did he spend time demonstrating his prowess at the line. Instead, he instructed them to focus on the back of the rim. St. Martin then proceeded to stand under the basket and watch their fingers. That is the instruction I remember so distinctly. My wife, Connie, and I later visited Ted and his wife, Barbara, and he told me the details of this experience. "I stood under the basket and watched to make sure their fingers were going toward the basket," he explained.

The Genius of Gabriele Wulf

Gabriele Wulf was trying to master the power jibe in windsurfing. This maneuver requires the well-timed coordination of leg and arm movements and appropriate weight shifts during its various phases, coupled with the precisely timed flipping the sail. Wulf focused on her foot placement and made little progress after a couple hours' practice. Frustrated, she decided to change her focus to the board's inclination. Focusing on the board led to success.

This experience in windsurfing was the inspiration for experiments she later conducted to examine the effects of external-based focus.[10] After numerous studies, the evidence is overwhelming. When executing any skill, she found, an external focus produces better results than an internal focus.

Internal Versus External Focus

Internal focus is when your attention is on your muscles and body parts; for example, elbow under the ball or fingers in the cookie jar.

External focus is when you focus your attention on the effect of the movement. In other words, it is on the environment around your body—for example, the flight of the ball or the back of the rim—and not your body itself.

The Seduction of Internal-based Instruction

Internal-focused instruction appeals to common sense. The shooting instructor is highly aware of what the player did wrong and knows precisely how it should be improved. The instructor provides this information to the player, who does their best to act on it. With a new self-awareness, the player uses this information while they practice. Sometimes this works. I have seen players practice and develop a better shot over time, which is super. But often, this approach fails, even with conscientious, hardworking players.

Dr. Wulf's Findings

Gabriele Wulf is a Professor in the Department of Kinesiology and Nutrition Sciences at UNLV. Her research has resulted in over 150 journal articles, as well as two books. Dr. Wulf studies factors that influence motor skill performance and learning; such as the performer's focus of attention and motivational variables. Wulf's studies have shown an external focus will:

- Make movements more efficient
- Improve retention
- Improve accuracy
- Increase speed
- Increase endurance
- Improve balance
- Improve maximum force production
- Reduce risk of cracking under pressure

An external focus has proven to work with athletes of all ages and in all sports.[11] It works for beginners as well as experts. Contrary to popular belief, beginners learn faster and improve more quickly when their attention is focused externally.

"When you adopt an external focus, you perform much more automatically and efficiently," Wulf says. "Somehow the body knows how to do what it has to do to achieve the desired outcome and that results in more fluid, efficient and accurate movements."

Your brain is better at directing your movements when you focus externally.

"When you ignore specific motions and pay attention to your surroundings, your form and performance will improve," Wulf says.

The main reason an inward focus seems to hinder performance is it constrains your brain, which locks in on one element. When you focus on body movements, you are consciously trying to control your actions. This internal focus disrupts the fluidity of the movement; you expend more energy than necessary and the accuracy of the movement declines. The more we think about our bodies, the less we are able to control them. Thinking about body parts and what they should be doing is detrimental to performance and learning.

A Real-life Case Study

Think back to the instruction Hopla and Palubinskas used. Hopla: "The shooting arm is fully extended with the elbow above the eyebrow and slightly in front of the body. It should stay in the shot line: the toe, knee, hip, shoulder, elbow, wrist, and hand all maintaining alignment." Palubinskas: "The index finger should be in the center of the ball and the shot pocket should be right in front of the eyes and two to three inches in front of the forehead."

Contrast that with Ted St. Martin's simple instruction: "Focus on the back of the rim."

The instruction which Hopla and Palubinskas provided focused their players' attention on their elbows and fingers and where they should be. This focus on body parts is an excellent example of internal-focused instruction.

Ted St. Martin instructed his players to focus on the back of the rim. The back of the rim is part of the surroundings and direct a player's attention outward. "Focus on the back of the rim," is an excellent example of external-focused instruction.

The detailed internal-focused instructions of Hopla and Palubinskas resulted in a decline in performance, while Ted St. Martin's external-focused instruction resulted in over a 20% improvement.

Personal Experience

Brynlee, my granddaughter, is 7. On an extended weekend visit, she resolved to learn how to ride a bike. She had been riding with training wheels, so this was going to be a milestone.

Brynlee made her first attempts on Sunday. I started to mention keeping her balance but caught myself and told her, "Keep the bike upright. If it starts to lean, turn the handlebars in the direction it's leaning." I was curious to see how well external instruction worked.

Brynlee fell numerous times, but she was persistent. Steering was a problem for her and a couple of times she ended up in the road ditch. But gradually she improved. I continued to give her the instruction, "Keep the bike upright."

Later in the evening, she made one attempt at riding. She did much better. She didn't fall and she steered around the sizable tree in our yard.

Brynlee hitting the open road.

The next morning Brynlee successfully navigated our yard, so we graduated to the dirt road by our house. After riding an eighth of a mile with no problem, we loaded her bike and drove to the park. She hopped on her bike and pedaled off like she had been riding for years. She cruised the loop around the park (which is a quarter mile) twice. After taking a break to play on the playground equipment, Brynlee announced, "I want to ride home."

'Home' was five miles away. "Are you sure?" I asked. She was. "Well, when you go out the front of the park, you need to take a right," I said. Off she went.

Brynlee didn't make it far. The first hill got her. "How about if you ride down the hills and I will pick you up and take you to the top of the next one?" I suggested.

That's what we did. I was extremely nervous, envisioning a severe wipeout going down a huge hill. Luckily, the first slopes weren't steep and no spills occurred. When we did get to the steeper hills, Brynlee was unfazed. On the back roads, she rode down every hillside without falling once. The distance she covered on her bike was over two miles. It was quite an accomplishment and Grandpa couldn't have been prouder.

Brynlee's progress was remarkable and I believe the external focus she employed was a factor. Focusing on keeping the bike upright allowed her to use all her resources to accomplish that task.

Paradigm Shift

The old paradigm of internal-based instruction says we have to get our body parts positioned correctly and every single body part has to move with precision.[11] In other words, the biomechanical movements of the body create the shooting motion. Your task is to position and move your body in exactly the right way as you execute the shooting motion. Shooting a basketball becomes a feat of biomechanical engineering to emulate the form of Klay Thompson. You accomplish this by endless practice, training your muscles to remember the exact position they should be in to execute the perfect shooting form.

The new paradigm of external focus holds that shooters should concentrate on what it takes to make the shot. Their body mechanics and movements occur naturally and instinctively in response to the player's intention and focus on making the basket. In other words, the biomechanics don't create the shooting motion; the shooting motion establishes the biomechanics called for to make the shot.

Great shooters aren't great shooters because they figured out what to do with their bodies. They are great shooters because they have learned how to apply force to the ball to make the shot and their bodies do whatever is necessary to achieve the end goal of a made basket.

Internal feedback phrases from the coach often result in a player performing worse than having no coaching at all. Studies have proven if you adopt an external focus when shooting, you will improve faster, be more efficient and reduce your chances of choking when the game is on the line.

How We Think Is Important

More than 30 years ago, Carol Dweck, Ph.D., author of *Mindset, The New Psychology of Success*, noticed her students differed in how they reacted to setbacks. Some rebounded, while others were devastated. Dweck coined the terms "*fixed mindset*" and "*growth mindset*" to describe the underlying beliefs students had about learning and intelligence.[12]

With a fixed mindset, students were concerned with appearing smart and refrained from taking on challenges in which they might fail. With a growth mindset, students understood they could become smarter and effort was required to accomplish this. Therefore, they looked for interesting and challenging situations which would lead them to higher achievement.

According to Dweck, the way teachers and parents praise kids can affect what mindset they adopt. Saying, "You are smart," influences a fixed mindset, which leads the student to question their abilities when they do struggle in school. To promote a growth mindset, Dweck advises parents to praise the child's process and strategies and tie those to the outcome. For example, "You kept at it and now you understand it"; "You put in the effort and look at how you've improved"; and "You studied more and you aced that test."

Moving Forward

Carol Dweck and Gabriele Wulf strike the same chord when it comes to learning. What their insights teach us is how we think and what we focus on affects our performance. Dialogue can help or hurt us. Wulf's studies on external focus provide insight into why the old way of teaching mechanics regularly failed. We can use this knowledge to our benefit.

However, the question which has consumed me for years is: Is it possible to use physics to help us make shots? And if so, how?

Chapter 2
Uncharted Waters

Learn a little about a lot but learn a lot about a little.

Anonymous

In my youth, basketball was my favorite sport. At that time, we didn't specialize in a single sport the way kids do today. We played the game of the season. Summer was baseball and winter was basketball.

First Book of Instruction

At the age of fourteen, the discovery of a detailed book on shooting a basketball was a revelation. While working at a neighboring school, I noticed a book in their library titled *Sharman on Basketball Shooting* by Bill Sharman, former Boston Celtic guard and arguably the best shooter of his era.[1] My boss, Mr. Robinson, allowed me to take it home. I studied it at length and hand-copied an entire chapter which provided detailed instruction on how to shoot.

Sharman opened my eyes to the fact there was more to shooting than picking up the ball and throwing it at the basket. I applied what he taught and my senior year I averaged 11 points per game and had a couple of games where I scored in the high 20s. However, this isn't as good as it sounds. We played in a small gym, which resulted in more possessions and higher scores. Everyone had decent numbers.

I did not attempt to play college ball, thinking I wasn't good enough. I did play recreationally and eventually took up coaching.

Become an Expert in Something

"Become an expert in something. Pick out one small area of something and become an expert in that area." This advice came from Joyce Brothers, Ph.D., who was the Dr. Phil of her time. She told the story of attending class and one of her professors giving students the advice, "Become an expert in something. Don't make it too general." The professor stressed it had to be something which intrigued them or they would get bored and move on to something else.

When I heard that advice, I knew I wanted to become a basketball shooting expert. Basketball shooting had intrigued me ever since reading Sharman's book. I resolved to learn everything I could about shooting and so began my quest for the perfect shot.

Learning Mechanics

I started by studying the shooting gurus. I bought every instructional video and book on the market. With the birth of the internet, more information became available, which I mined as well.

After several years it became apparent to me the gurus didn't have the answer. Shooting instructors all emphasized mechanics; building the shot from the ground up, starting with the stance and ending with the follow-through. Each had their take on what it took to be successful and they were right—their method did work. But never to the level of consistency I desired.

Learning to Coach

During this time, I started coaching at small schools in northeast Kansas. I took great pride in my ability to teach the mechanics of shooting to my players. Some it helped, but others it didn't.

I was extremely fortunate my friend and mentor, Orville Dodson was kind enough to take me under his wing and teach me how to coach—or I should say, he tried. Orville has over 40 years experience teaching and coaching and he knows basketball. Orville influenced me in many ways, but the most important thing he taught me was how to break things down to a level the players could understand. Orville is an absolute master at that, primarily because his depth of knowledge is incredible. He knows more about basketball than anyone I know.

Orville has a wry wit. The first year we were coaching together, we were watching the kids run through their warm-ups before the game. "You know," he said. "I have been standing here watching and the only difference between our team and their team is when we shoot a 15-foot shot it doesn't go in and when they do, it does."

Halfway through my third year, Orville commented, "I looked at the last 30 years at this school and the average tenure of the head coach in basketball is 2.3 years. You can quit anytime."

On another occasion, during pregame warmups, I knew we were in trouble. We were playing a team we had beaten earlier in the season and it was apparent our players expected to win. Confidence is necessary, but overconfidence is not. Our players were obviously overconfident and I was worried. Running through warm-ups, they were lackadaisical and distracted. Wondering how to deal with their lack of focus, I asked Orville. "What did you do when your kids showed up not ready to play?" Expecting some valuable words of wisdom, I waited expectantly for his response.

Orville just smiled wryly and said, "Lose."

The Catalyst to a Different Approach

After 20 years of studying shooting, I had plateaued. I had tried everything. I felt like I had been led down a bunch of blind alleys and experienced a ton of frustration in the process.

As I read John Fontanella's book *The Physics of Basketball*, it was as if someone flicked on a light switch. It altered my perspective. Fontanella used physics to solve the distance aspect of shooting. This approach was different. What I loved about Fontanella's approach is he used science to obtain his answer. He didn't rely on opinions. He applied physics. The laws of physics are ever present and consistent. I love that. Also, using physics shifts your focus away from *you* to what is happening to the ball, which is what is essential. Ultimately, how and where you apply force to the ball determines where the ball is going to go. Mechanics become secondary.

Fontanella had solved the long/short aspect of shooting. He composed an equation involving the four factors which affect the flight of the ball: gravity, drag force, Magnus force and buoyancy of the ball. For the slowest moving ball when it nears the rim (which gives you a shooter's touch), Professor Fontanella calculated when shooting a free throw, a 6-foot player should launch the ball with a 51-degree launch angle. A

7-foot player, releasing from a higher vantage point, should use a 48.7-degree launch angle.[2]

How do you argue with science?

Fontanella taught me something else, too. He pinpointed the release as the most important aspect of shooting. Therefore, which fingers leave the ball last is critical. For him, that was his middle and ring fingers, which was different than what the experts taught. I tested it and found it worked quite well. Learning of an entirely different shooting method caused me to question everything I had previously learned about shooting a basketball. It was a watershed moment.

Since Fontanella had the answer for distance, I resolved to solve the deviation aspect of shooting.

First, I went out to my shop and built a crude stand with an arm and a protractor, which allowed me to measure degrees. I would stand this device by a player at the free throw line, set it at the appropriate angle for the player's height and have them shoot while watching to see if their launch angle was correct. Usually, they could see it out of the corner of their eye and readily match the angle.

Changing to a Physics-based Approach

I changed how I taught shooting to players. I switched from teaching mechanics to introducing options. I was amazed at the response. Players were interested, engaged and intrigued. They were more involved and had more ownership of their shot. It turned shooting into a learning experience for them. Plus, it created an atmosphere of working together instead of following directions. I became a believer in this approach and resolved to give up teaching mechanics.

A Pick-me-up Comment

Connie and I traveled to Dallas from our home in Kansas to meet Gary Boren, free-throw-shooting coach for the Dallas Mavericks. Gary gave us a tour of the facility and we discussed shooting at length. I showed him a few of the props I used when working with players and the launch-angle device was one of them. Before we left, Gary casually said, "As far as knowing something about shooting, of all the people I have talked to, you are in the top 1 to 2%."

"Would you mind repeating that?" I responded.

I wanted Connie to hear his comment. Connie has been my most avid supporter, but by this time she was questioning the amount of time I was investing in basketball. I had completed my third year in the role of head coach at Axtell High School in northeast Kansas. That season I had implemented the original three-point shooting style of Grinnell College's David Arsenault. This radical barrage of three-point shots was something the parents and superintendent questioned. We finished third in the state in three-point shots made but only won half of our games. If you think this might be a recipe for a coaching change, you are right. They let me go when the season was over. "We are taking a new direction" is the phrase a fraternity of coaches have heard and now I had joined the club. It hurt to get fired. Connie and I loved that group of boys and leaving was a disappointment. However, I have no regrets. The kids enjoyed that style of play and so did I.

Gary Boren's comment was timely because my self-esteem was at a low point. It elevated my spirits and inspired me to stick with it. (Thank you, Gary Boren!)

The next morning, Gary called and asked if I would sell him a "launch-angle device." We agreed on a price of $100 and when I returned home got busy building him one. When I finished, it cost me $40 in materials and $60 to ship. Not exactly what you would call an astute business deal. However, it spawned the idea I might be able to market that device as a shooting aid. We contacted Advanced Manufacturing Institute (AMI), an affiliate of Kansas State University to help improve it.

Making a Shooting Video

Dale Wunderlich and Brian Rempe were the project managers assigned to help us. They were immediately impressed with my knowledge of shooting and suggested creating an instructional shooting DVD, rather than trying to market a niche item. We shifted gears and established that as our project goal.

AMI helped us find CVP Productions and we met with Rob Fanning, who agreed to shoot, direct and edit the video. I wrote the script. CVP had a screenwriter revise it and add some corny jokes. The material is mine; the jokes are not.

We hired child actor, Jacob Hayes to be the lead and deliver the information. Instructional videos tend to be somewhat dull and fail to hold viewer's attention. Jacob solved that problem. He was entertaining.

The video is called *Secrets of Shooting*. I introduced the concept of teaching more than one way to shoot, among other things. Back in 2008, this was a radical concept. No one else had ever done that. Traditionally, shooting instructors taught their way and the masses were to mimic their example. I opened the door to experimentation, which turns shooting into more of a learning experience.

Connie believed in me and backed this project 100%. We invested more money into the video than I care to admit. From a financial standpoint, it was a disappointment. We have sold videos all over the world—Italy, Japan, Malaysia, Germany, Canada, Australia, and France—but not enough to recoup our costs. From a cynical standpoint, our timing was impeccable because YouTube was catching on. Why buy a video when you can watch something online for free?

The video did open doors. We met Ted St. Martin, who has the Guinness World Record for most consecutive free throws (5,221) and Dr. Tom Amberry, who at the age of 71, made 2,750 free throws in a row. Both these men encouraged me to start shooting, with the intent to try for a record. I resisted. Their encouragement wasn't enough to get me to the gym.

The Inspiration to Start Practicing

To do that it took Daniel Coyle's book *The Talent Code: Talent Isn't Born. It's Grown. Here's How.* Coyle introduced me to myelin. Myelin is a white, fatty substance which surrounds the axon of some nerve cells, forming an electrically insulating layer.[3] Myelin acts as an insulating substance which helps the brain's circuitry get the signal to your muscles faster and more accurately. You build myelin through practice.

What inspired me to start practicing was the knowledge that over time, I would be changing my brain. However, I doubted I would break any record. When I was in my 20s, after reading about Ted St. Martin, I went to the gym to see how many I could make in a row. I quickly dispensed with that idea as my accuracy was no match for Ted.

Coyle's book also brought to my attention the work of K. Anders Ericsson, who is one of the world's leading experts on experts. Ericsson has devoted his life to the study of what it takes to become great. He has determined the key is what he calls *deliberate practice*. Deliberate practice is practicing in such a manner that you are persistently stretching the outer limits of your ability.[4]

Does This Stuff Work?

I was curious to see if Coyle and Ericsson's insights worked. I had my doubts but was willing to give it a try. I had concerns because Coyle had pointed out myelin growth slows as we age. According to his book, it might not grow after the age of 50. Well, I was 52. I am happy to report you can increase myelin when you are older. In fact, it is now widely known you can develop it throughout your life.

Growing Talent

After a month of going to the gym regularly, I noticed a significant improvement in my accuracy. After two months I was consistently making 100 in a row. I checked out records and applied on Guinness World Records™ website to break the "most free throws in one minute with one ball and one rebounder." The current record was 25.

About this time, my father was diagnosed with bladder cancer. We met with the doctor and received a death sentence—according to him, if Dad took chemotherapy, he might last a month. Without it, he only had a couple of weeks. Dad opted to bypass the chemo and went home. Dad truly enjoyed the following week because he had all his kids around. He passed away on Nov. 9, 2009.

During this time, I stopped going to the gym. It was a Saturday when I went back. I made 246 in a row, missed one and then made 200 in a row before missing again—446 out of 448 shots. After that day, I stopped working on accuracy and started focusing on speed. Before long I made 25 in a minute during a practice session with a coaching friend, Ben Scism, rebounding.

Shortly after this, I heard back from Guinness. There must have been some mistake because the application for the record stated, "Most free throws in a minute is 48 and is held by David Bergstrom from Sweden." I immediately fired back an email asking for the one-ball, one-rebounder record.

Speed Shooting with Multiple Balls

They didn't respond. However, it soon occurred to me that 48 in a minute wasn't out of reach. With a rebounder, it took one second for the ball to get to the rim and another second to get it back from the rebounder. With unlimited balls, I wouldn't have to wait for the ball to return. Theoretically, 25 in one minute with a rebounder would equate to 50 in a minute with more shots.

A piece of cake, right? How hard could it be? I soon found out. There is a speed/accuracy trade-off and I had overlooked the fatigue factor. Fire 50 balls up for a minute and see how well you do in the last 10.

Not to be deterred, I built a crude makeshift rack which held multiple balls. I put eight on it and asked Connie to time me with a stopwatch. I fired them up, turned to her and asked. "How long did that take?"

"5.6 seconds."

"Great!" I replied. "That's faster than one per second."

"But you didn't make *any* of them."

That didn't faze me. I had confidence my accuracy would improve because, by this time, I believed in K. Anders Ericsson's deliberate practice concept. I could sense myself getting better because I was noticing subtle nuances I hadn't before. I had access to more information while I shot.

Shooting Coach

Ryan Noel asked if I would be the shooting coach for his girls team at Valley Heights High School in northeast Kansas. He lobbied the school board to issue me a contract, to comply with Kansas High School Athletic Association (KSHSAA) guidelines. KSHSAA prohibits anyone from working with players during the season on a voluntary basis—you must be a part of the coaching staff, which means the school board has to issue you a contract. The school board agreed to hire me for the sum of $1 (which came to 93 cents after taxes), with the stipulation I would work with the boys team as well. I agreed and I was in the money.

Noel was on board with my unorthodox teaching techniques and the girls enjoyed a successful season. The team agreed to help with a record attempt on a Saturday. The school was closed on Thursday when a snowstorm hit the area. Determined to practice, Connie and I drove the 30 miles through snow-packed roads to the gym. In hindsight, it was pretty stupid. Fortunately, we made it there and back home safely.

My First Guinness World Record

Saturday morning, Jan. 10, 2010, with Connie coordinating the logistics, Ryan Noel handing me the balls, and the Valley Heights girls basketball team rebounding, I made 50 free throws in one minute to surpass Bergstrom's record of 48. Connie assembled the required documentation and mailed it to Guinness. They approved it and I joined the ranks of Guinness World Record holders.

Breaking this record received some local publicity and before long I started practicing to surpass the two-minute record of 68, held by Rick Rosser, one of the world's best free throw shooters.

More Guinness World Records

In March, I broke the two-minute record with 88. Next up was the 10-minute record of 298. In June, I broke it (336) and established a 30-second record (33) as well.

This became a pattern. I would set my sights on one, go after it, break it and then proceed to the next one. Learning to shoot blindfolded, underhanded and with my left hand took some time, but by then I had learned how to improve.

To date, I have set or broken 22 Guinness World Records and I am the current holder of 14 GWR in different categories of free-throw shooting.

Learning Biomechanics

How was I able to do this? By drawing knowledge from unconventional resources. After reading Fontanella's book, I started studying physics and biomechanics. In 2009, Connie slipped on the ice and broke her leg. Her recovery involved two surgeries and numerous doctor appointments. Since I was her caretaker, I was regularly present when she met with the orthopedic doctors. I would wait until Connie was through asking questions and when the doctor asked, "Do you have any more questions?" I would jump in with mine.

"I have one, Doc. Is the hammering motion of the wrist more stable and consistent than the dart-thrower's motion?" Or, "How much variance is there between people when it comes to the bucket-carrying angle of the elbow?"

The first time I did this, Connie looked at me as if I had lost my mind. Then she got mad. "Why did you ask him that? Aren't you concerned about my leg?"

She had covered all the questions about her leg I had. Picking a doctor's brain was my chance to learn something which might help my shooting. After the first time, Connie came to expect it. When the doctor asked his final question, she would look at me expectantly and say, "Well, here's your chance." (She would tell people and laugh about it later.)

One doctor was kind enough to lend me two of his textbooks, which I studied at length.

Scientific Studies

Joan Vickers' book, *Perception, Cognition and Decision Training* covered her work on the *Quiet Eye* concept, the gaze which underlies higher level of skill and performance during sporting events.[5] Vickers' work provides help to free-throw shooters.

In her book, I learned of Dr. Raôul R.D. Oudejans of Amsterdam University and his 2005 study on improving the visual perception of jump shooters. His experiment resulted in an 18% improvement in three-point accuracy during game conditions.[6] I contacted Oudejans via email and he was kind enough to send me other studies he had done. I was still coaching at the time and Oudejans was interested in learning of the practical application of his scientific research. I guess I was one of the few crazy enough to put it into practice.

Unorthodox Sources of Knowledge

After discovering this new information, I became enamored with studies and broadened my search for clues. I pored through hundreds of studies trying to glean information relevant to shooting. I emailed two authors of these studies and they were kind enough to respond with answers to my questions.

Did I understand everything I read? No. Some of it was over my head. However, it was interesting and occasionally I uncovered some pertinent information.

I learned we all have movement variability between throws; it is a question of how much. According to these studies, the strongest finger is the middle finger and the most powerful position of the wrist is 7-degree ulnar deviation.[7] Deviating from this position can result in a loss of power of 20 to 30%. Also, the dart-thrower's motion is the most natural movement of the wrist. This action consists of a movement from extension and radial deviation to a position of flexion and ulnar deviation.[8] Activities which demand control and maximum grip strength utilize this movement.

What did this have to do with shooting a basketball? Although I didn't know it at the time, I was laying a foundation for the development of a physics-based approach.

Broadening the Search

Articles and books provided helpful insights as well. In 2011, The New Yorker ran an article titled *Personal Best: Top Athletes and Singers Have Coaches. Should You?* This article was by Atul Gawande, M.D., a surgeon, who asked Robert Osteen, M.D., a retired general surgeon, to be his coach.[9] Osteen was to watch Gawande perform an operation and critique him.

Part of the criticism was several times during the surgery; his right elbow rose to shoulder level and sometimes higher. "You cannot achieve precision with your elbow in the air," Osteen told him. "A surgeon's elbow should be loose and down by his side."

If a surgeon loses precision with his elbow shoulder level or higher, wouldn't a basketball player lack control as well?

(LevanteMedia/Shutterstock.com.)
Note the position of his elbow. Starting his shot from a lower position would provide more control.

Sensitive Palm

In the book *What the Dog Saw* by Malcolm Gladwell, there is a quote by Marc Goldstein, a sensory psychophysicist: "The finger has hundreds of sensors per square centimeter. There is nothing in the science of technology that has even come close to the sensitivity of the human finger with respect to the range of stimuli it can pick up. It's a brilliant instrument. But we simply don't trust our tactile sense as much as our visual sense."[10]

Next time you're in the gym, shoot a few free throws with your eyes closed. You will be more aware of the weight of the ball and you will feel the ball leave your fingertips better. I guarantee it.

Bucket-carrying Angle of the Elbow

Another tidbit of useful information concerned the "bucket-carrying angle of the elbow." "This 'carrying angle' of the elbow allows your forearms to clear your hips when you swing your arms while you walk. It also helps us carry buckets without them hitting our legs. What was interesting is females have a more significant carrying angle to their forearms than males.[11] So, theoretically, when shooting, shouldn't a female's forearm be angled more than a male's?

(Photo Works/Shutterstock.com)　　　　(Richard Paul Kane/Shutterstock.com)

Above and Beyond

From *The Art of Learning* by Josh Waitzkin, I learned when you are fighting a martial arts expert, don't blink. Here is how he describes it: "On his blink, or just before it begins, I pulse into a one-two combination, left, right, into his body. … This can happen before he finishes blinking. He goes flying onto the ground and comes up confused. … Afterward, opponents have come over to me and implied that I did something mystical. They were standing and then on the ground and they didn't feel or see anything occur in-between."[12]

What does this have to do with shooting a basketball? Nothing. It merely demonstrates there is a higher level of expertise you can attain. Whether you are punching during the time your opponent blinks or focusing on the release when shooting, you can develop a higher level of proficiency with the right training.

Overcoming Institutional Bias

We all are biased. Our past influences and experiences create filters which information must pass through. The more experienced and established we become in an institution's line of thought, the more entrenched we become in the institutional mindset. We lose our ability to be creative thinkers. In other words, we lose our ability to think outside of the box. Once we reach this point, we see what we look for and we look for what we know.

Here's a personal example. Celtics legend Larry Bird and legendary coach Red Auerbach made an excellent instructional basketball video. In it, Bird states, "As I bring the ball up, my elbow is in, my wrist is bent back and I push off with *my middle two fingers* and follow all the way through."[13] This statement contradicts the index finger release that I had learned. For some reason, my mind glossed over that comment and it didn't compute. I watched this video several times and that statement didn't register.

It did compute when John Fontanella said it on an NPR program. Answering a question, Fontanella stated, "What is important is the release and which fingers leave the ball last. And for me, that is my middle and ring fingers."[14]

That finally clicked. I heard and computed that because I was listening with rapt attention. I had been studying *The Physics of Basketball*, and was intrigued. Also, I was in the process of questioning conventional thinking.

When facing evidence that is different than what we know to be true, we tend to ignore it. Once the evidence becomes overwhelming, we make the shift mentally. Evidently, this is what I did.

Bias Slows Progress

Stanford's Hank Luisetti changed the game in the 1930s by shooting with one hand instead of two. Hall-of-Fame Coach Nat Holman represented the establishment's thinking of the time when he said, "That's not basketball. If my boys ever shot one-handed, I'd quit coaching."[15] Even five years later, the St. John's coach at the time, Joe Lapchick, said, "I can't

(Pavel Shchegolev/Shutterstock.com)
We are different. The concepts presented in this book allow all players to be successful.

be persuaded that two on the ball doesn't make for far superior shot control and a greater percentage of hits." Only after years of watching more one-hand shooters emerge did coaches change their thinking. Luisetti is now considered a great innovator.

A more recent example of institutional bias is displayed in the movie *Moneyball*. The old-timers thought baseball players had to look the part. Oakland Athletics Manager Billy Beane's only concern was whether the player could get on base. Beane's line was a classic: "We aren't selling jeans here. Can he *hit?*"[16]

Initially, institutional bias was a problem for me. A little voice in my head was saying, "This is not proper mechanics. You aren't supposed to be doing it this way." Once I earned several Guinness World Records using my new methodology, I told that voice to get lost.

"Every shot should look the same" is an example of institutional bias in basketball. The robotic approach, or the mechanical approach; is what I learned. For years, this is what I taught. The old standardized, one-size-fits-all approach. It never occurred to me there might be a better way to teach shooting. I now know there is.

Shooting Breakdown

In summer 2016, I hired Landon Lucas, a senior at the University of Kansas, to visit my shooting clinic to entice kids to attend. Lucas observed my PowerPoint presentation and participated in on-court exercises. Lucas taught the kids how to dunk and throw alley-oops. In closing, I asked Lucas if he had anything he wanted to add. "I have been playing basketball now for over 20 years, at a very high level, with great teams and coaches," he said. "But I have never seen a breakdown of shooting to the extent you have seen here today."

Shoot Better than the Pros

In today's National Basketball Association, players display high levels of shooting skill. Stephen Curry, Klay Thompson, Kevin Durant, Kyle Lowry, Kawhi Leonard, J.J. Redick, and Kyrie Irving are extraordinary. However, you will also be able to attain skill levels which are off the charts with this book's concepts. Understanding and implementing this methodology will sail you into uncharted waters. Physics provides a true north. Shooting is an optimization problem and what makes it difficult is there are many ways you can shoot effectively. However, there are common denominators which link different methods and I will be sharing them with you.

Mission Accomplished

I am pleased to have accomplished my original goal, which was to understand shooting. After years of chasing my tail, going down blind alleys and being mystified at why I could never achieve the skill level I desired, I now know how and why the shot goes in—and why it doesn't.

    I am going to share this knowledge with you. This knowledge has transformed my shooting and more importantly, will dramatically improve yours. Once you understand the concepts presented here, you will never again experience the frustration of missing shots and not knowing why. The next chapters will explore how you can best use physics and biomechanics to increase your shooting percentage. The player with a basic understanding of physics who effectively brings it into their game will outshoot and outscore every other person on the floor.

Chapter 3
The Secret of the Release

"Any fool can know. The point is to understand."
Albert Einstein

Great shooting is the ultimate equalizer. I hope what I tell you in the following pages will give you the groundwork to become a great shooter.
As you probably know, field goal percentage is the dominant factor in winning basketball games.[1] In 2014, the San Antonio Spurs' record field-goal percentage of 52.8% was the best in the NBA Finals history. It is not a coincidence they won the championship, defeating the Miami Heat.[2]

It is common for teams to win due to a hot shooting night. In 2015, Pittsburgh defeated North Carolina 89-76. North Carolina shot 49%, but Pittsburg shot 65%.[3] After the game, North Carolina's Brice Johnson said, "We were scoring, but when they are not missing a shot, it's hard. We are a really good rebounding team, but if it's going through the net, there is nothing you can do."

Let me say right up front there is no *best* way to shoot for everyone. Each person has individual physical and mental characteristics which can be used to their advantage. The way you shoot can be unique, personal and distinct. While there isn't one single way to be a great shooter, I have discovered specific fundamental concepts great shooters share.

Rather than teaching one particular method, I am going to explain critical concepts which will improve your shooting. I will also introduce the *Centerline Concept*, the most accurate way of launching the ball straight through the center of the basket.

Pro players such as Klay Thompson, Kyrie Irving, Stephen Curry, James Harden, Kevin Durant, Kyle Lowry and others are raising the bar and showing what is possible. We will shine a light on what they are doing which makes them so successful.

A Mind-blowing Concept

I mentioned this earlier but here is the full story. Some years ago, I heard John Fontanella, author of *The Physics of Basketball,* interviewed on NPR and he said something which blew my mind. So I picked up the phone and called him.

"Professor, I heard you say you shoot with your middle and ring finger leaving the ball last," I said. "Did I hear you correctly?"

"Yes," he answered. "That is correct."

"But I was taught to shoot with my index finger being the last finger to leave the ball. I wasn't even aware you could shoot that way."

"Well, I started shooting that way back in college," he answered. "During my freshman year, I hurt my index finger and I experimented a bit and found I could shoot better releasing off my middle and ring fingers. So that is how I shot from then on."

I paused a moment, digesting his words. Fontanella's book was groundbreaking and I had a great deal of respect for him. The fact that one year he had led the country with a free-throw average of 92.2% (in the National Association of Intercollegiate Athletics) added credibility as well. Might this be the missing link to the secret of shooting?

"Well," I said. "It is disheartening to think for 49 years; I have been shooting the ball the wrong way."

I will never forget Fontanella's reply.

"It is not the way you were shooting was *wrong*. It is just shooting that way is not very biomechanically friendly."

Time for a Different Approach

That conversation shook my beliefs about what I had learned about shooting a basketball. Was Fontanella right? Biomechanics is how we move. Was it possible I had been using a release which wasn't natural my entire life? What else did I believe to be true that wasn't? What other misconceptions did I harbor?

Searching for the Perfect Shot

This conversation motivated me to research and experiment with different ways to release the ball. I relentlessly re-examined every aspect of shooting fundamentals. It has been a long but insightful journey. What I have discovered is more valuable than I ever could have hoped. It is an entirely different approach to shooting a basketball which is easy to understand and works for everyone. This new approach is grounded in physics rather than opinions. It expands the limits of what is possible. If you come to understand it fully, you will realize it is a real game-changer.

How It Began

As I mentioned earlier, what started my journey was the advice, "Become an expert in something." Shooting a basketball had always intrigued me. At that instant, I knew I wanted to become an expert in basketball shooting.

Searching, Searching, and More Searching

All the shooting gurus I studied stressed mechanics. Building a strong foundation of fundamentals, starting with your stance and finishing with your follow-through. These experts advocated shooting with your index finger *or* your index and middle finger being the last fingers to leave the ball. Using this method proved to be difficult for many of the players I coached. It did help some, but others it did not. A couple of players stand out in my mind, who worked and worked but failed to achieve any level of consistency. Personally, I was decent but never great. I was unable to attain the high level of ability of the gurus I was imitating. Something was missing.

Elusive Answers

If you go to YouTube, you can find endless information on how to shoot a basketball. The problem is, not all of it is valuable. How much time do you have to sort out what is helpful and what is dubious? What works and what doesn't? What do you do when an idea presented works one day and doesn't the next? Misinformation is worse than no information because it leads you down a dead-end street.

If you were to listen to ten different shooting experts, you would get ten different takes on how to shoot. Much of the information will be contradictory. How do you know who is right and who is wrong?

Some years ago, my sister-in-law was visiting with her granddaughter, who enjoyed basketball. I popped in an instructional shooting video by a well-known shooting expert and while we watched it, I provided commentary: "He is right on-target with this point." "That is not relevant." "That is wrong and this is why..."

After the fourth time I corrected his material, my sister-in-law looked at me. "What would I do if you weren't sitting here telling us what is good or bad? It all sounds good to me."

Excellent question. How do you separate truth from rubbish?

Eureka! Insights from Professor Fontanella

Something resonated when I read Fontanella's book, *The Physics of Basketball*. In it, he broke down ball flight into four factors: gravity, drag force, Magnus force and buoyancy of the ball. He put them in an equation and calculated the ideal arc for the slowest moving ball when it nears the rim.

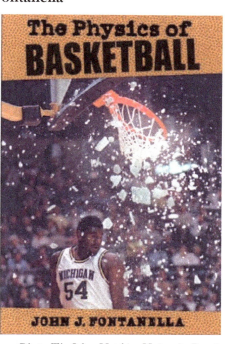

(Photo: The Johns Hopkins University Press)

The Catalyst to a Different Approach

Fontanella's approach appealed to me because he used science instead of his personal viewpoint to solve the problem. Everyone has an opinion on how to shoot. I wanted to *know* I was right and be able to prove it, in the same manner as Fontanella did.

He used a physics-based approach to explain the distance aspect of shooting. Since Fontanella had solved the long/short factor of shooting a basketball, I thought I could use a similar approach to figure out missing to the left or right. My problem was I didn't know physics or biomechanics—which, I surmised, were both vital to analyze in reaching my goal.

I began studying physics, anatomy, and biomechanics. I would like to say I have become an expert in these areas, but that would be a reach. However, I did learn the basics as they relate to shooting a basketball.

Many Variables

There are a significant number of variables involved in shooting a basketball. At one point, I decided to list them all but quit after 48. There are probably more. The existence of numerous variables creates a problem. Variables which interact in subtle and complicated ways become difficult to understand. To compound this issue, the more advanced the player, the more nuanced the variables may be.

Investigating Biomechanics

Biomechanics is the study of how we move.

The cornerstone of shooting a basketball with consistency is to take your body's natural motions and make the most efficient use of them. Your physical makeup provides unique opportunities to customize your shot to take advantage of your strengths.

On my journey to become an expert in basketball shooting, I studied and experimented at length with numerous options. Through experimentation came insights. I documented them and kept track of what worked best in achieving the highest consistency in getting the ball through the hoop.

The Physics of Shooting is Constant

Although an individual's variables are involved, the physics of shooting is constant and works all the time. Science provides guiding principles which help you separate truth from rubbish.

What is the Most Important Aspect of Shooting?

If you tie a ball on a string and swing it around your head, it will go in a circle. The moment you let go, the ball will immediately take off in a straight line.

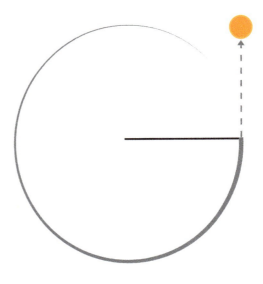

(Diagram by Garrett Steinlage)

If you are trying to control the direction of the ball, the moment you let go is critical. So it is with shooting a basketball. The moment of release is *the* crucial moment. Where you are applying force to the ball *at the moment of release* will determine the direction of the ball's flight.

For this reason, shooting a basketball should be taught backward. Start with the release first. The moment the ball leaves your fingertips is the most important aspect of shooting. Everything else is secondary. If you get the release right, you will have success. Focus on the most critical element first. Then move on to secondary issues which might be hurting consistency.

To Do Is to Know, Not to Do Is to Guess

In September 2009, at age 52, I started shooting every day. My objective was pretty simple; to see how good I could become. On November 9, I made 246 free throws in a row, missed one and then made 200 in a row. At that point, I stopped going for accuracy and switched to speed shooting.

My goal became two shots per second, which is much faster than most players shoot. Eventually, I came close to achieving this by making ten free throws in 5.04 seconds. To understand how fast this is, visualize

(Photo courtesy of Jake Crowther)
During the filming of Inside Edition. It takes approximately one second for the ball to get to the rim. Notice there are four balls in the air at the same time.

one ball going through the net with the next one in flight halfway to the rim. Since it takes approximately one second for the ball to get to the basket, this would represent a pace of two shots per second.

Missing Left

While speed shooting, I noticed my shots tended to stray to the left. I would make the first three or four shots and then miss to the left repeatedly. It was a problem I couldn't seem to get past. With my elbow in line with my side, my wrist's natural movement is inward and to the left (actually 17 degrees, according to my measurement). Shooting slow, I could get my fingers moving toward the basket but when I started shooting faster this natural inward movement would kick in and I would miss left. It wasn't "muscle memory." It was how my wrist typically moved. In essence, I was trying to teach my wrist a new movement of going *at* the basket, which evidently didn't follow the natural pulling direction of my wrist muscles.[4]

I knew I had a problem and I wasn't willing to wait years to fix it. Rather than invest the time teaching my wrist to move a new way, I decided it would be better to use the natural movement of my wrist and build my shot around that.

Adjust Your Shot for Your Own Biomechanics

I countered my tendency to miss to the left by moving my left foot forward and my right foot back. Although this is the polar opposite of what most players do, this position squared my shoulders to the basket, which caused my fingers to move straight toward the basket when I snapped my wrist. Reversing my stance solved the problem. Putting my left foot forward turned my shoulders, so the natural movement of my wrist sent the ball at the basket no matter how fast I shot.

I thought this was unconventional and unique until I read *The Paradox of the Free Throw*, a dissertation written by Jim Poteet.[5] In it, Poteet mentioned there was only one coach who advocated shooting with the left foot slightly forward: John Wooden.

At that point in my development, this strategy worked well for me. I set my first four world records using this stance. However, I no longer use it. I have learned better solutions which negate the tendency to miss to the left.

During this time, I was experimenting with different options to decide which finger should leave the ball last. I tried shooting with my thumb, little finger and everything in between. They all worked; some

(Diagram by Garrett Steinlage)
Shooting is an optimization process.

better than others. But with each release, it was relatively easy to reach a high level of proficiency.

The Ah-ha Moment

As I experimented, I kept looking for conceptual links between methods. What did they have in common?

This search continued for years. I eventually realized we have *more than one way* to apply force to shoot the ball straight to the basket. Overlooked through the years is the fact that we have another option available to us to send the ball straight.

Two Methods of Shooting

Every shot you see in a game falls into one of two categories. These two groups are what I call Straight-line Thrust and Centerline.

1. The Straight-line Thrust Method

The most-often implemented release is the Straight-line Thrust Method. This is a straight-line thrust of the hand and fingers toward the basket and is what everyone is trying to achieve. Players try to keep their fingers moving towards the rim while they shoot and follow-through toward the basket.

(Pavel Shchegolev/Shutterstock.com)
Index finger release--note the alignment and how he uses his thumb and middle finger to direct the ball to his index finger.

2. The Centerline Method

A second method I have identified is the Centerline Method.

First, I define "centerline" as an imaginary line which runs through the center of the ball and the middle of the basket.

(Diagram by Justin Smith)
A vertical line running through the exact center of the ball and the exact center of the basket is called centerline.

The critical element of the Centerline Method is the application of force through the centerline of the ball and basket, *at the moment of release*. It doesn't matter how you do this, or what fingers you use. When you control the application of force through the exact center of the ball at release, the ball will go straight. This action is simple physics.

Here is the surprising part. Any direction of movement toward centerline will work as long as you are at the exact center of the ball at release. Every shot does not have to look the same when you use the Centerline Method. Being able to send the ball straight while your hand moves at an angle to the basket was the aspect that was difficult for me to wrap my head around. Never has that been taught. The focus is on applying force through the center of the ball at release. This allowance for movement variability raises the

(Aspen Photo/Shutterstock.com)
Middle finger release—note the middle finger in the exact center of the ball at the moment of release.

bar of what players can do. With this method, higher levels of skill are possible.

The Centerline Method: A Pool-shooting Analogy

A valuable way to help you visualize the Centerline Method is by looking at the game of pool. If you want to sink a ball which is directly in line with the cue ball and the corner pocket, you hit the cue ball straight into the object ball.

However, if the cue ball isn't in line with the object ball, you must execute a cut shot. The key to making a cut shot is to have your cue ball contact the exact center of the object ball in relationship to the corner pocket. A technique for teaching novice pool players where the cue ball should hit the object ball is to ignore the cue ball and pretend you are going to hit the object ball directly into the pocket. The center of the object ball relative to the corner pocket is where the cue ball needs to contact. This location is the centerline of the ball and the corner pocket.

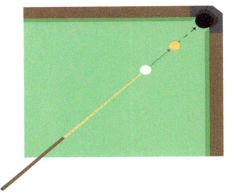

(Diagram by Garrett Steinlage)

When shooting a basketball, the same principle applies. Your hand acts in the same manner as the cue ball. In billiards, the moment of impact determines where the ball will go. In shooting, it is the moment of release. Where you are applying force at the moment of release defines the path the ball will travel.

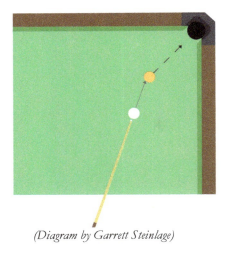

(Diagram by Garrett Steinlage)

The Straight-line Thrust Method: Making It Better

With the Straight-line Thrust Method, the emphasis is on alignment. To best execute Straight-line Thrust, you should be stationary and every shot should look the same. (An NBA player who does this well is Kyle Korver.) Although you can align your toe, knee, hip, shoulder and elbow using a slightly staggered stance with the shooting foot forward, it isn't necessary. You *can* do all this. But what causes the ball to go straight is your wrist.

(Diagram by Garrett Steinlage)
The pivot point of your wrist.

When you shoot, the wrist snap happens last. The release is the most important aspect of shooting and occurs during the wrist snap. Therefore, the wrist snap becomes vitally important to the direction the ball travels.

If you keep your fingertips moving *in line* or moving *parallel to* the pivot point of your wrist and the center of the basket, the ball will go straight.

With the Straight-line Thrust Method, your fingers do not have to be in the exact center of the ball.

Many people think the elbow makes the ball go straight. This joint acts like a hinge and therefore can be used to dictate the initial direction toward the basket. But the wrist snap trumps the movement of the elbow because it happens later.

(Diagram by Garrett Steinlage)

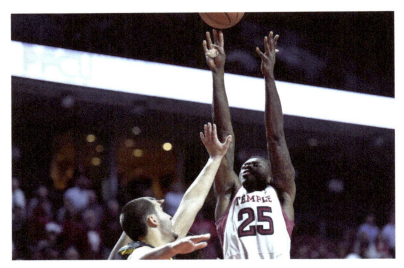

((Aspen Photo/Shutterstock.com)
An example of a straight-line shot. Note the fingers are not in the exact center of the ball.

(Aspen Photo/Shutterstock.com)
Note the extension of his elbow but the wrist snap has yet to start.

The Straight-line Thrust Method: About That Elbow Angle

Incidentally, the forearm should not be precisely vertical when shooting because we have a "bucket-carrying angle" in our elbow which

(Aspen Photo/Shutterstock.com)   (Day Owl/Shutterstock.com)
Note the angle of their forearms. Women have a greater "carrying-angle" of their elbow than men do. This angle varies from player to player.

creates a slight angle to the forearm.[5] In a study by the National Center for Biotechnology Information, the carrying angle in females was more significant than the angle in males (16.2° ± 3.2° versus 13.6° ± 3.0°).[6] Therefore, for a more natural wrist motion toward the basket, your forearm will be angled inward slightly. For most players, keeping the forearm perfectly vertical when shooting is biomechanically awkward and therefore ineffective. A slight inward angle to the forearm helps achieve more consistent shooting success.

The Straight-line Thrust Method: Release Finger

With the Straight-line Thrust Method, the index finger has historically been the finger of choice to leave the ball last. In theory, keeping your elbow under the ball and your index finger in line with the rim should result in a straight shot. However, when you use the Straight-line Thrust Method, any primary control finger or fingers will work. It is interesting to note many of today's best shooters use the middle-finger release. Watch Stephen Curry, James Harden, Klay Thompson, Kevin Durant, Kyle Lowry,

Kawhi Leonard, J.J. Redick and Kyrie Irving in action. The basketball leaves their middle finger last.

Ultimately, if you keep your fingertips in line or moving parallel to the pivot point of your wrist and the center of the basket, the ball will go straight. The ball invariably travels straight when your fingers move straight toward the basket at the moment of release.

Occasionally, you see players arc their fingers outward during their follow-through. When they do this, they must get the timing of the release right. If their fingers are moving toward the basket at the moment of release, the ball will go straight.

(Marcos Mesa Sam Wordley/Shutterstock.com)
Note how James Harden uses all his fingers and thumb to control his shot. The ball started on his palm and moved to his fingertips. What happens during the next split-second will determine whether he makes or misses this shot.

However, if they are a split-second late, the ball will miss to the right.

The Straight-line Thrust Method: The Other Four Fingers

Even though our focus has been on which finger leaves the ball last, *all* fingers should stay on the ball as long as possible.

Why? For better control. Use your thumb and four fingers to control the direction of the ball. Some players even close their fingers as the ball leaves. Closing their fingers as the ball leaves keep their fingertips on the ball a split-second longer, which provides more control.

The Straight-line Thrust Method: Finger Placement

A noteworthy aspect of the Straight-line Thrust Method is finger placement on the ball is primarily irrelevant. Your fingers don't have to be in the center of the ball. The ball will go straight if your fingers are providing a straight-line thrust toward the basket. Whether the fingers are in the middle or slightly to the side of the basketball makes no difference when you maintain control of the ball. The key is the straight-line movement of the fingers moving parallel to the pivot point of the wrist and the basket.

Matching one's fingers to the curvature of the ball helps you have a consistent, natural wrist movement. However, when using this technique, it is imperative your fingers go straight toward the basket. Although it is rare to see this, a few players utilize this method with success.

(Leonard Zhukovsky/Shutterstock.com)
DeAndre Jordan is using a shooting method which lacks control over the center of the ball at the moment of release.

The Straight-line Thrust Method: Caveat

While it seems to make sense to keep your index finger in line with the rim on release, it is difficult to do consistently. Though this teaching has been around for eons, players rarely master the Straight-line Thrust Method with the release point being the index finger because this is a learned movement rather than a natural one. In other words, the Straight-line Thrust Method is not biomechanically friendly. In general, it is what everyone is trying to do. However, the other option you have available is the Centerline Method. This method will more likely help you reach your full potential as a player.

Introducing the Centerline Method

The Centerline Method basis is Newton's Second Law of Motion: The acceleration of an object as produced by a net force is directly proportional to the magnitude of the net force, in the same direction as the net force and inversely proportional to the mass of the object. As it pertains to shooting, it would be fair to say you can summarize Newton's Second Law by saying, "The ball goes where we direct it."

(A. Ricardo/Shutterstock.com)
The beauty of the Centerline concept is it allows you to make unorthodox shots. If his middle finger reaches the center of the ball at release, this ball will go straight.

Apply force on the right side of the ball while you are releasing it and you will miss left. Apply force on the left side and the ball will err to the right. When you control the center of the ball relative to the center of the basket, the ball will always go straight.

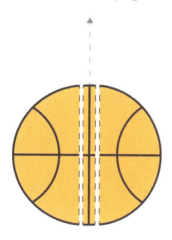

(Diagram by Garrett Steinlage)

The beauty of the Centerline Method is your focus is on one variable: controlling the centerline of the ball at the moment of release. It doesn't matter if you are standing square to the basket or at an angle. It doesn't matter if the ball is sitting on your palm or fingers. You don't have to keep your hand/fingers going straight toward the basket while you shoot. It doesn't matter if your hand moves straight or curves while it is moving forward. Which fingers you choose to leave the ball last is entirely up to you, because it doesn't matter. *All* releases work when you control the center of the ball relative to the center of the basket at the moment of release. The Centerline Method tops everything.

How do you know where the centerline of the ball is when you shoot? It is the high point on the ball. The ball is a sphere. Therefore, the highest point of the circumference will be centerline. You can develop your sense of touch to perceive the curvature of the ball. With practice, you can expand the sensitivity in your fingertips. Great shooters acquire a feel for the ball. It is a requirement for being an outstanding shooter.

(Aspen Photo/Shutterstock.com)
Frank Mason was fun to watch. His mastery of the Centerline Method was unprecedented at the college level.

You can develop your touch for the center of the ball and when you do, you will get to a point where it is rare to miss. Since the basis of this concept is grounded in physics, this is a "can't miss" shot. Every shot you shoot that you control the center of the ball relative to the basket is going straight. Physics dictate that to be true.

The Centerline Method: Proven Success on the Court

A player who uses this centerline concept to its full advantage is Frank Mason III. Kansas guard Mason won the 2016-17 Wooden Award and was recognized as the nation's most outstanding player. Mason also secured Naismith and Associated Press Player of the Year honors during a standout senior year with the Jayhawks.[7] In

(Aspen Photo/Shutterstock.com)
Frank Mason was fun to watch. His mastery of the Centerline Method was unprecedented at the college level.

(Muzsy/Shutterstock.com)   (Aspen Photo/Shutterstock.com)
A finger roll is when a player shoots with one hand and lifts his fingers to the centerline of the ball, rolling the ball into the basket.

36 games, he averaged 20.9 points, while shooting 49.0 percent from the field and an astounding 47.1 percent from the three-point line.

At the college level, Mason was the best I have ever seen at utilizing the Centerline Method.

At the pro level, Jerry West was one of the best at utilizing the Centerline Method. Although West had a standard technique he used extensively, he could also use other methods to score, which may be why he made so many clutch shots over his career. Stephen Curry is a current NBA player who uses the Centerline Method to his benefit. Curry utilizes this concept to make an impressive variety of shots—runner, scoop, layup, etc. His performance is an exceptional example of what is possible with this technique. Once you genuinely develop Centerline understanding and utilization, you become virtually impossible to guard.

Notable NBA greats used this technique in their repertoire. Elgin Baylor, Oscar Robertson, Wilt Chamberlain, George Gervin and Julius Erving were the pioneers. Although they used the Centerline Method with proficiency, it is highly unlikely they understood why it worked. A basic understanding of this concept will help anyone become a better scorer.

The Centerline Method: Shooting Close to the Basket

Shots close to the basket regularly use the Centerline Method. Lebron James is a master of this technique. He uses the Centerline Method to convert a wide variety of shots around the basket. Another superb example is Hall of Famer George "The Iceman" Gervin's finger-roll, a shot he softly launched from any angle.

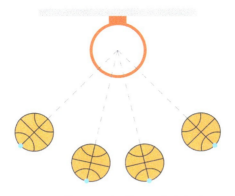

(Diagram by Garrett Steinlage)
The most important aspect of the Centerline Method is to control the centerline of the ball and basket at the moment of release.

The Centerline Method: Distance Shooting

The Centerline Method works for long-distance shots as well. Kevin Durant has perfected controlling the center of the ball when he shoots from outside. Controlling centerline provides him with both accuracy and versatility.

(Leonard Zhukovsky/Shutterstock.com)
Note how Kevin Durant starts with his hand on the side of the ball. The slight turn of his hand serves to center the ball over his palm.

Centerline Method Feedback

The first time I experienced shockingly consistent success with the Centerline Method, it was surreal. I was making everything. It was motivating. I started pushing the limits; launching shots from different positions. When I nailed centerline at release, the ball went into the basket. Plus, I could feel the centerline with my fingertips. I knew the instant the ball left my hand when it was going to miss. I could feel the weight of the ball shift slightly and the miss would correlate to which way the weight shifted. Knowing what went wrong provided the information I needed to correct on the next shot. It eliminated the mystery of shooting.

The Centerline Method: Finger Placement at Release is Optional

To send the ball straight your finger placement at release can be the same but *doesn't have to be*. Over time, you may notice the corresponding contact areas of your fingertips which are leaving the ball in the same place. You may start trying to hit these contact areas because this seems to help you make shots—which it does. Nailing the exact contact area of your fingertips at release works exceptionally well. But don't lose sight of the fact you are controlling the center of the ball in relation to the exact center of the basket. That is key. Your fingers can be in any position on the ball as long as you do that.

In other words, you have two options here. You can pinpoint the contact areas of your fingertips at release, *or* you can learn to feel the curvature of the ball and control the center of the ball with whatever fingers are positioned to do so. Either way, the moment of release—the instant the ball leaves your fingertips, is the critical moment. Get the release right and you are money.

The Centerline Method: Finding the Right Arc

The centerline for propelling the ball straight to the target never varies. Centerline is permanently the exact center of the ball in relationship to the exact center of the basket. But how do you attain the right arc?

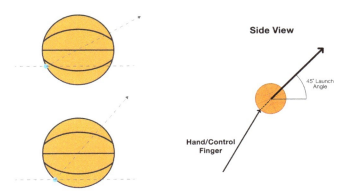

(Diagrams by Garrett Steinlage)
Where your fingertips leave the ball at release varies due to distance. These two balls illustrate the positions of a long-distance shot and close to the basket. The diagram on the right shows the spot for a three-point attempt.

Your distance from the basket will determine where you should be applying force when the ball leaves your fingertips. The spot your fingers separate from the ball should correlate to the arc required for your distance from the rim.

A two-foot shot requires a launch angle of approximately 72-degrees.[8] The starting point of a line through the center of the ball at a 72-degree angle would signify where your fingertips should be at the moment of release.

For long shots, this spot would be higher on the backside of the ball. A three-point shot would be at about a 45-degree angle.[8] But there is an ideal spot which corresponds to your distance from the basket. When you hit the back-iron of the rim and the ball bounces out, at the moment of release your fingertips were too high on the ball. Fingers too high on the backside of the ball at release will result in a flatter shot. Focusing on what area of your fingertips are in contact with the ball at release will provide feedback, which will help you improve.

Advantages of Using the Centerline Method

1. The Centerline Method increases movement variability. You are no longer required to be robotic. We aren't robotic by nature and trying to do the same action over and over is difficult, if not impossible. This approach allows freedom of movement, which will result in additional opportunities to score.

2. You have immediate visual feedback using the Centerline Method. By watching where the ball goes, you will instantly know what you did and how to correct it.

3. External focus is better than internal focus. Internal focus is all about you (i.e., how you stand, where you should put your elbow, etc.). External focus is focusing on what you need to do (i.e., apply force through the exact center of the ball). Studies have repeatedly proved external focus results in more improvement and better performance during game conditions.[9] The Centerline Method encourages the use of external focus.

4. You will improve faster with more biomechanical options using the Centerline Method. Research has proven mixing up the material results in deeper learning.[10] Variable practice is better than shooting the same shot 1,000 times.

5. Shooting will be a more enjoyable experience. Knowing it is physics-based provides a reduction in self-awareness and self-evaluation. Your confidence can handle misses better when you understand what happened. You missed centerline of the ball and basket at the moment of release. Next time, nail centerline and you will make it.

6. A positive learning environment comes with the Centerline Method. It enables players rather than inhibits them. Players have more ownership of their shots.

7. The Centerline Method doesn't place restrictions on players the way the teaching of mechanics does. Fewer restrictions allow players more room for growth. They can stretch the outer limits of their boundaries, rather than be locked into some preconceived notion of what they can or cannot do. The Centerline Method raises the bar for what you can do. You will be surprised at how good you can become.

8. Results are immediate. You don't have to wait two to six weeks to train and develop a new movement to your muscles or conform to some idealized shooting form. You can have success today if you execute the Centerline Method of controlling the center of the ball at release.

9. The Centerline Method allows individuality. Each player is unique. We have differences and this approach will enable you to use your strengths to find and use what works best for you.

How to Become Good at Anything

There are three ingredients to becoming a skilled shooter (or skilled at anything): knowledge, practice and time.[11] To become good at anything, you have to learn everything you possibly can, train at the edge of your abilities and persist.

Knowledge

Accurate and timely knowledge is essential. To streamline and accelerate a player's progress, knowledge and understanding of core concepts will make a phenomenal difference in shooting success.

Focused Practice

Repetition is essential but mindless repetition can be fruitless. Some players put in the time and are erratic shooters. The essential requirements of effective practice are *focus* and *feedback*.

Focused practice is concentrating and being in the moment. Feedback is understanding the cause of mistakes and making adjustments to fix them. Feedback is imperative for improvement.

Time

Nobody becomes proficient at anything in one day. Everyone who has ever achieved significant success has put in the time to do so. Persistence is an essential quality to becoming proficient in anything. Stay with it. Stay with it. Stay with it.

Changing My Teaching Methodology

I remember the first time I provided a player with options, rather than dictating one particular style. The response was overwhelmingly positive. The player was interested and engaged at a high level. Shooting a basketball had become a *learning* experience, rather than a direction-following robotic performance. That session ended my days of teaching mechanics.

The days of teaching one form to everyone should be over. Excellent basketball shooting instructors are like hairdressers; they know one style doesn't work for everyone. The truth of the matter is you can become proficient in a variety of ways. Selecting your best method should be your decision.

How to Teach the Centerline Method to Kids

The Centerline Method is a better way to teach shooting. Providing physics-based instruction opens doors to learning and enhances creativity.

Imagine you are addressing a bright-eyed bunch of youngsters looking for guidance.

"Kids, there is an imaginary line which runs from the center of the ball to the center of the basket. Let's call this centerline. When you apply force through the exact center of the ball when it comes off your fingers, the ball will go straight. If your fingers don't quite get to the center of the ball, you will miss to the side. Watch the ball and see where it goes to know what you did. When it goes straight, you know you hit exact center. Now let's get to shooting!"

This approach will help kids achieve success much more quickly than telling young players, "Stand like this, grip the ball this way, get your elbow in, etc. ..." Teaching an idealized perfect-shooting form is setting kids up for failure. Young players quickly disengage and become frustrated with this approach. Kids love to see the ball going through the hoop. Help them accomplish this and they will be motivated to keep practicing.

Instead of starting with mechanics, teach shooting backward. Communicate what is essential from a pure physics-based perspective and address the release before anything else. Let the player operate without unnecessary constraints. It is interesting how often players adopt the secondary issues when they focus on what is most important. Get the release right and the other elements of shooting fall in line. Other shooting fundamentals may or may not help you be more consistent, but when you get the release right, it makes up for a multitude of sins.

What is important is to help kids learn *why* they made or missed the shot. Remind players to pay attention to where the ball hits on the

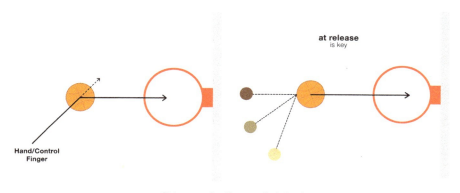

(Diagrams by Garrett Steinlage)

rim. "The ball goes where it does because of where you were applying force at the moment of release." A deep understanding of this simple concept along with focused practice will allow you to outshoot everyone you play.

Making the Shot

Watching where the ball goes and knowing what you must do to make the next shot is feedback. Feedback is vital for improvement. With the mechanical approach, players miss and generally have no idea what caused the miss. The Centerline Method provides accurate information so you will improve faster.

To Improve, Push Your Limits

Try speed shooting for insights into your body's natural movements. Shooting slowly allows you to develop bad habits because you can make virtually any technique work. Speed shooting will show you your natural tendencies and help you make your shot more efficient. You may find evidence to help you make real improvements. Incorporating some speed shooting into your practice will help you improve faster. Plus, it's fun.

Eyes on the Target

It is common to see players look up at the ball while it is in flight to the basket. Due to my high school coaching experience, I am not a fan of looking up to admire the ball for a couple of reasons:

1. The speed of the game is faster than in practice. In game situations, when players shoot under pressure, they will look up slightly quicker than usual, which results in their shot falling short.

2. Players who do this sometimes tilt their heads back when they look up, which affects their body balance and causes the ball to miss.

Other sports stress the need not to follow the ball in flight. While making short putts in golf, it's best to keep your head down and listen to the ball going into the hole rather than nervously looking to see if you've made it. If you look at high-speed photos of tennis legend Roger Federer after he's hit the ball, he is still looking at where the ball was, not admiring his shot. So, although I don't have any studies to back me up on this point, it has been my experience that keeping your eyes on the target during the entire shot reduces the tendency to miss short in game situations.

The Wrist and the Importance of the Wrist Snap

The wrist has five muscles which enable it to move in any direction. Each muscle has a preferred pulling direction. When the desired movement doesn't fall along the path of this preferred pulling direction, the closest two or three muscles work together to create tension and resist each other to produce the desired movement. The muscles of each person's wrist vary in a range of motion and strength.

The bending and extending motion of the wrist when shooting a basketball is called the wrist snap.

In 1996, Roger Miller and Stuart Bartlett conducted the study *Contributions to Ball Speed on a Free Throw*. In this experiment, their subjects were 15- to 19-year-old boys. They analyzed the amount of force each body part contributed to a free throw. According to their study, 59% of the force to the ball came from the wrist snap.[12] Contrary to popular belief, the arms and shoulders contributed more than the legs (81% upper body compared to 19% lower body).

Your wrist snap should bring your control fingers to the center of the ball at release for a straight shot every time. It also should be natural. Why? Because it will increase your range. You will be generating more power with your wrist. A natural wrist snap puts your wrist in its strongest position. Furthermore, it will provide more consistency. A natural wrist snap is a universal trait among great shooters.

(Aspen Photo/Shutterstock.com)
This finger position is an excellent example of the Centerline Method. Note his middle finger is in the exact center of the ball at the moment of release.

Pinpointing the ideal contact areas on your fingertips at release may help you use your natural wrist motion more efficiently.

In Summary

There are two methods to apply force to the ball: Straight-line Thrust and Centerline.

What is the fundamental difference between these two?

The Straight-line Method is based on the *basket* and directing a straight-line movement of your hand and fingers toward it. The Centerline Method is concerned with the *ball* and where you apply force to it.

Success in the Straight-line Method relies on how well you align your thrust to the basket. It helps to be stationary so you can direct your focus on the basket and shoot toward it. The adage, "Make every shot look the same" is a sound idea. The less deviation, the better.

Success in the Centerline Method is based on your ability to control the center of the ball at the moment of release. When you achieve this, the ball goes straight all the time.

Your options expand when you focus on the ball. With the Centerline Method, you can have movement variability. All shots don't have to look the same. Also, lateral movement is less of an issue because you are

(*Leonard Zhukovsky/Shutterstock.com*)   (*Leonard Zhukovsky/Shutterstock.com*)
Note the different follow-throughs for these excellent shooters who use the middle-finger release.

dealing with the *ball* and where you apply force to it. (When moving laterally, you will have to compensate for momentum, but we will get into how to do that in later chapters.)

Controlling the exact center of the ball relative to the basket allows you to react and adjust to the demands of the game. What route your fingertips take to get to the middle of the ball does not matter. What is important is they are in the exact center of the ball at release.

Physics dictate when you apply force through the center of the ball relative to the middle of the basket at the moment of release, the ball will go straight. Whether it is the moment of impact in the game of pool or the moment the ball leaves your fingertips, the physics is the same. The ball reacts to the application of force. Where you are applying pressure *at the moment of release* will determine whether the ball goes straight or not.

This concept is extremely important and bears repeating—the moment of release is the most critical element of shooting. Controlling the center of the ball, relative to the basket at the moment of release, is the single most crucial element of shooting a basketball. Accomplish this task and the ball will go straight every time.

Game-changing Concepts

- The release is the most important aspect of shooting.

- The cornerstone of shooting with expertise is to take your body's natural motions and make the most efficient use of them. Your physical makeup provides you exclusive opportunities to customize your shot to take advantage of your strengths.

- Teach to the individual. The days of teaching one style to everyone should be over. There is no one best way of shooting which works for everyone.

- You can pinpoint the contact areas of your fingertips at release, or you can learn to control the center of the ball with whatever fingers are positioned to do so.

- Use external focus. There is no need for internal-focused instruction (for example, "Keep your elbow in"). Internal-focus can be detrimental to improvement.

- Variables are different for each of us. Using physics along with individual biomechanical strengths typically works best.

- Knowledge is an essential ingredient for improvement, as is practicing. They are both necessary to reach your potential. Mindless repetition is treading water. Practice alone will not get it done and knowledge without practice won't either. Both are necessary for improvement. Focused and engaged practice will move you forward.

- Rather than advocating one particular method of shooting, it is better to teach the concept that the release is what is essential. Where you apply force at the moment of release is the critical element.

- Visualize and focus on the centerline of the ball and basket. The integral component of the Centerline Method is to control the center of the ball in relation to the middle of the basket. By applying force through the center of the ball at the moment of release, you will be sending the ball straight every single time, no matter how you shoot.

- All shots do not have to look the same. Yes, this is an entirely different approach which requires a different mindset, but if you come to understand it fully, you will see the opportunity to increase your potential and gain a competitive advantage over your opponents.

- Accurate feedback will help you improve faster. Watching where the ball goes provides information. This information is vital for making adjustments and continued improvement.

- For maximum consistency, it is helpful to use a natural wrist snap.

Conclusion

We've all experienced days when we are *on*. When you have such a day, it is because you are accurately applying force through the center of the ball at the moment of release. Physics and biomechanics in sync—what a game-changer!

Exceptional shooting is a great equalizer. Hopefully, the concepts discussed here will provide the groundwork for you to improve your shot. There are levels of skill in shooting a basketball which is as yet unrealized. Players can attain higher levels of expertise. You can be one of those players.

Part 2

The Physics of the Free Throw
by
Larry M. Silverberg, Professor of Mechanical & Aerospace Engineering
North Carolina State University, Raleigh, North Carolina

Dr. Larry M. Silverberg is a professor of mechanical and aerospace engineering at North Carolina State University (1984 to present). His work lies in the areas of theoretical and applied dynamics.

Along theoretical lines: One of the challenges that faces engineering is with dynamic modelling that integrates disciplines (mechanical, electrical, and thermodynamic), scale (small and large), and phase (gaseous, liquid, and solid states). His monograph entitled *Unified Field Theory for the Engineer and the Applied Scientist* (Wiley VCH, 2009) is a treatment of modern field theory with the goal of making recent developments in unification more accessible to engineers and scientists. Toward making this subject more accessible, he also teaches a graduate-level course entitled *Modern Modelling*, which focuses on modern computational issues of integrative modelling.

Along more applied lines: Dr. Silverberg has published over fifty journal articles on applied dynamics topics and two applied books. In the nineteen eighties and nineteen nineties, he pioneered new methods in the active control of aerospace structures; engineering developers all over the world now use his method of uniform damping control of spacecraft. In the nineteen nineties, he developed a method of calibrating ground-based antennas to enable them to detect objects in space. Based on this work, he led the design and fabrication of the Orbiter Ejector, flown on three unclassified shuttle flights (and other classified flights). The Orbiter Ejector released test spheres into orbit for the calibration of a worldwide network of hundreds of ground-based radar stations. The calibration enabled the military, for the first time, to detect incoming ballistic missiles in support of President Reagan's Stars Wars program.

In the area of aerial systems, Dr. Silverberg has led the development of several new unmanned aerial vehicle systems, including one vehicle that swims and flies and another one that drops from a fighter aircraft. His research team has also developed several path-planning algorithms for unmanned aerial systems. They developed and flight-tested the, *Central Command Architecture for Systems of Large Numbers of Autonomously Operated Aerial Vehicles*. They also developed the first autonomous algorithms that enabled UAS to locate and circle thermals (leading to several world records).

His research team is currently developing an autonomous path-planning algorithm for small-class unmanned aerial vehicles. It produces lumbered flight, that is, the lazy motion that we see birds of prey (about 5-25 pounds) use to conserve energy. Also, his team is currently developing autonomous algorithms for aerobatic maneuvers, and a new autonomous architecture for autonomous payload changing of UAS. Such agencies as NSF, NASA, DARPA, and Northrop Grumman support Dr. Silverberg's research team.

Outside of work, Dr. Silverberg has always enjoyed basketball. He began to study the dynamics of the basketball shot some twenty years ago. The work began with the development of a suite of simulation tools that accurately predicts the trajectory of a basketball, its collisions with the backboard and rim, and statistically analyze the probability of a shot. He and his colleagues applied the techniques to study best practices in the free throw and the bank shot. Current work in basketball focuses on the ageless question of the superior gender when it comes to the free throw shot.

Special Chapter

Introduction

When I first read Bob Fisher's manuscript and found that it was physics-based, I was thrilled. Not just because of my work in physics and engineering, but because I had already come to believe that physics is the foundation of good shooting and this was at center stage in Bob's approach.

I am a dynamicist. That means that I study the motion of bodies. I study dynamics from a fundamental perspective, trying to understand the underlying reasons behind the motion of anything. In my day job, I scrutinize the underlying laws that govern the motion of bodies at the macroscale. The great scientists of the past discovered them by studying the patterns found in macroscale behavior without understanding from where the patterns originate. Historically, paradoxes arose in the basic laws of macroscale science and my work contributes to resolving those paradoxes. Of course, the basketball shot is a dynamics problem, too, and I love the game of basketball, so I have also spent bits of time over the years studying the dynamics of the basketball shot (probably more time than I can justify). This chapter reflects the pleasure that I have derived from studying the physics of the basketball shot.

Long ago, I had grown to appreciate how important a precise mathematical model of a dynamical process can be to understanding the process's behavior. With a mathematical model, the scientist can play the "What if" game. What would happen if I were to do this or that? I can use the mathematical model, if constructed to find a "best" solution, to study a problem in detail toward gaining insight into the nature of the problem. Of course, sports is full of dynamics problems and I could not help but look at the dynamics of the basketball shot.

I began to study the dynamics of the basketball shot more than twenty years ago. My colleague, Dr. Chau Tran, and I decided to develop an accurate method of simulating the trajectory of a basketball accounting for all of the ways the ball can strike the hoop and the backboard. In 2003, we wrote the paper entitled *Numerical Analysis of the Basketball Shot* and published it in the world's premier mechanical engineering journal. It was the first paper to show how one can accurately and quickly simulate any basketball shot and run millions of cases to get probabilities of success. Whereas, in practice, it would take you quite a while to take millions of shots, we let the computer do this and it gave us an ability to play the "What if" game and objectively study best practices. The results I describe in this chapter result from those simulations.

The first basketball shot that my colleagues and I studied was the free throw. We studied it between 2001 and 2006. Most of the results in this chapter come from that period. In 2008, we published the paper entitled *Optimal Release Conditions for the Free Throw in Men's Basketball*. It appeared in the *Journal of Sports Sciences*, the premier journal in the science of sports. After studying the free throw, we went on to studying the bank shot. We studied it between 2006 and 2012. That work led to a surprising discovery—that the best bank shots strike the rim on a V-shaped line on the backboard and that the planes of all of their trajectories intersect on a focal line behind the backboard—leading to the development of a training tool for the bank shot called the backboard-V. In this chapter, however, we will focus on the free throw.

How to think about the free throw as a dynamics problem

To start out with, I would like to ask you to think about the free throw as a dynamics problem. I am asking you to take a step backwards while reading this chapter. I will *not* go into how it is best to hold the ball, how to adapt your shot when you are moving, or how to shoot a floater. Indeed, in this chapter I will not focus on the details that explain how to achieve the best shot. Instead, I will discuss the physics behind the basketball shot. What do all of these shots have in common? How do forces create motion? What are the best basketball trajectories? By going back to the basics, if the basics are truly complete, you will grow to appreciate the common physics behind all of the shots and the true meaning behind all of the terms you will be using when thinking and learning about all of the shots.

Let us start by considering that every shot has two stages. There is the pre-launch stage when the ball is still in your hands and there is the post-launch stage when the ball is flying through the air heading to the basket. Of course, the physics of each of these stages is very different. During the pre-launch stage, you have all of the control in the world over the motion of the ball while during the post launch stage you have no control. All you can do is wait and see whether the ball enters the basket. Therefore, as a player, your mission is clear. Your goal during prelaunch is to impart the best launch conditions possible. These launch conditions are the parameters that describe the motion of the ball at the instant that the ball leaves your fingertips and begins on its way to the basket. All of your effort comes down to launch conditions at a single instant of time.

The pre-launch stage and the post-launch stage are different in another relevant way. It is a matter of perspective. From the point of view of the pre-launch stage, the launch conditions occur at the end of the pre-launch. They are final conditions. On the other hand, from the point of view of the post-launch stage, the launch conditions occur at the beginning of the post-launch stage. They are initial conditions. This is significant because the pre-launch stage, being that the launch conditions are final conditions, and being that the individual controls the ball during this stage and that each individual is different, it follows that there are many different "best" ways to get a set of launch conditions. Your body has many ways it can move to give the ball the same launch speed, launch angle, etcetera, and each body does this best differently. On the other hand, the post-launch stage is very different. During the post-launch stage, being that the launch conditions are initial conditions, being that a basketball has a unique set of properties, and being that the ball follows only one fateful path once the launch conditions have been imparted, it follows that there is just one set of "best" launch conditions. Putting the pre-launch stage and the post-launch stage together, there are two objectives:

(1) Associated with the pre-launch stage: achieving the best body motion, recognizing that there is a multitude of best body motions.

(2) Associated with the post-launch conditions: achieving the

best ball motion, recognizing that there is essentially just one best ball motion.

The first objective of achieving the best body motion, that which leaves the ball with the best launch conditions, is about consistency. It is about learning what the best body motion is and how to repeat that motion consistently. It takes lots of practice. The second objective is very different. It is not about conditioning your body as much as it is about conditioning your mind. It is about knowing what the best ball motion is. This is important, too. Many years ago, I conducted a computer simulation that compared two shooters who shoot the ball at the same level of consistency. I gave the first player a rather flat shot and gave the second player the right amount of arc. The consistency that I gave the players was such that the first shooter achieved a 70% free throw. The second player, who I gave the same level of consistency, turned out to achieve a 90% free throw. The point is that shooting the best trajectory is critical.

> From the dynamics viewpoint, the overall objective in the basketball shot is to achieve the best launch conditions with the greatest consistency.

Before addressing what the best launch conditions are, let us first ask ourselves a more basic question. What are the launch conditions? The answer is that there are three sets of launch conditions. There are launch positions, launch velocities, and launch angular velocities (See figure of Launch Conditions).

The figure shows the launch point of the center of a ball. It is quite general in that the ball is located at an arbitrary point above the floor. In the free throw, we refer to the center of the ball's path as it flies through the air as its **trajectory**. At the launch point, the ball has three **position components** labeled x, y, and z. The figure shows the x-y-z coordinate system with its origin placed at the center of the hoop. The x component is the x coordinate of the position of the launch point measured from the origin of the coordinate system. The y component is the y coordinate and the z component is the z coordinate.

Next, at the launch point, the ball has three **velocity components**. We can represent them in a variety of ways. For example, we could consider its components in the x, y, and z directions, referring to them as the

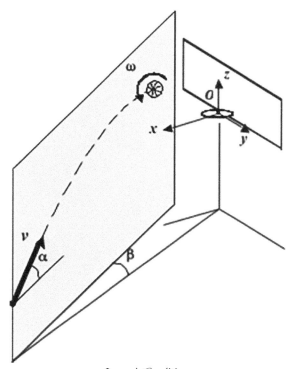

Launch Conditions

rectangular components. Here, we will *not* use rectangular components. Instead, we will use its spherical components. The three spherical components of the velocity are its magnitude v, its **forward (pitch) angle** α (called alpha) and its **side angle** β (called beta). We more often refer to the magnitude of the velocity as its **speed**.

Finally, at the launch point, we give the ball **angular velocity components**. The angular velocity components are the rates at which the ball rotates about its three axis. Here, we use its rectangular components, that is, the rates at which it rotates about the x-axis, the y-axis, and the z-axis. We denote the angular velocity components by ω_x, ω_y, and ω_z (called omega-x, omega-y, and omega-z). In the game of basketball, we more often refer to the ball's angular velocity as its **backspin** or just its **spin**. When shooting a free throw, the ball spins predominantly about its y-axis and you launch the ball with the top of the ball spinning toward you. If we are to be precise, though, the ball also has a small x-component and z-component of spin.

In summary, the three sets of launch conditions are:

$$\left.\begin{array}{c} x \\ y \\ z \end{array}\right\} \text{position coordinates}$$

$$\left.\begin{array}{c} v \\ \alpha \\ \beta \end{array}\right\} \text{velocity coordinates}$$

$$\left.\begin{array}{c} \omega_x \\ \omega_y \\ \omega_z \end{array}\right\} \text{angular velocity coordinates}$$

Newton's Laws of Motion

Let me now explain to you how we impart motion through forces. This is the subject of classical mechanics. It starts with Sir Isaac Newton's First, Second, and Third Laws of Motion and his Universal Law of Gravitation (dating back to 1687). Although beyond the purposes of this book, those of you who have philosophical leanings will find it fascinating that Newton, when introducing his laws, asked the scientific community to take them on faith, lacking at that time any explanation behind his own laws. Scientists eventually unraveled the mystery, but it took centuries. It was not until about 1970 when scientists figured out that the laws are part of the broader "universal model" of modern science, which interprets science merely as an agreed-upon representation of reality that has certain practical rules but that claims no absolute truths about reality. Putting the philosophy of science aside, we express the classical tenants of science in terms of Newton's Laws (See figure of Newton's Laws) as follows:

> **Law 1:** A particle moves at a constant speed and direction in absence of a force f acting on it.

Law 2: A particle accelerates in proportion to the force f acting on it. We shall refer to the proportionality constant as the particle's mass m.

Law 3: Two particles interact with equal and opposite forces that act along the line between them.

Law of Gravitation: The interaction forces between two particles are attractive, inversely proportional to the square of the distance D between them, and proportional to the product of their masses. We refer to the proportionality constant as the universal gravitational constant G.

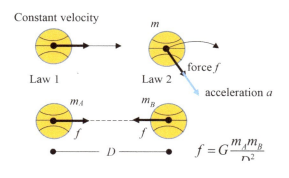

Newton's Laws

I now ask that you read over Newton's Laws. In fact, you may want to read them over several times. Then, consider the following:

(1) Newton required that we take all of the measurements in a frame of reference that is stationary or moving at a constant speed and direction. We refer to that frame as the **inertial frame**. In my past work, when simulating the motion of a basketball, I always took my inertial frame to be the x-y-z coordinate system whose origin is located at the center of the basket (see figure Launch Conditions).

(2) Notice that all of Newton's Laws refer to the objects as particles. Moreover, notice that he never considered the sizes of the particles in any of the laws (although he considered their

masses). It is important to be aware of this because it means that the sizes do not play any role in his fundamental laws. How can that be? I will explain this below.

(3) Newton's First Law (top left in figure Newton's Laws) is a special case of his Second Law (top right in figure Newton's Laws). According to Newton's Second Law, when the particle's acceleration is zero the force acting on the particles is zero, too. Furthermore, we know that acceleration is the time rate of change of velocity. That is the definition of acceleration. Therefore, when the acceleration is zero, the time rate of change of the particle's velocity is zero. In other words, the particle's velocity is not changing, or stated differently, it is constant. These are precisely the conditions stated in Newton's First Law.

(4) Newton's Third Law is about how any two particles interact. It is how we can build up from two particles to any number of particles and create bodies that have size and shape. In classical mechanics, bodies are thought of as systems of particles.

Although Newton's Laws do not mention bodies, we can apply Newton's laws to each of the particles that make up a body and derive from Newton's laws a new set of laws that pertain to bodies. When we do this, we get three new laws that look exactly like Newton's first three laws except that we can replace everywhere the word "particle" with the word "body." The force acting on the body now acts at the body's mass center and the velocity and acceleration are of the mass center, too. When the body is a basketball, the mass center is the same as the ball's geometric center. We call the three new laws: **Law 1 for the Translation of a Body, Law 2 for the Translation of a Body, Law 3 for the Translation of a Body.**

These three new laws still mention nothing about a body's sizes and shape but by being applicable to bodies, we can at least consider how they apply to basketballs and basketball players.

In addition to these three new laws pertaining to the translation of a body, there is a fourth law that arises from manipulating Newton's first three laws. It is a new law about how a body rotates. We can state it as follows (see figure for Moment and Force):

Law 2 for the Rotation of a Body: A body undergoes angular acceleration in proportion to the moment acting on it. We shall refer to the proportionality constant as the body's rotational mass.

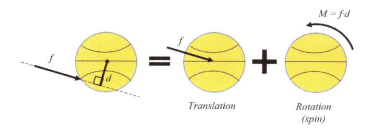

Moment and Force

This new law requires some explanation for I have not yet told you what a moment is. The explanation is as follows:

(1) A moment is like a "rotational force." Whereas a force causes a body to undergo an acceleration, a rotational force causes a body to undergo an angular acceleration. The analogy is between translation and rotation. Length, velocity, acceleration, and force are translational quantities and angle (angular length), spin (angular velocity), angular acceleration, and moment (angular force) are corresponding rotational quantities.

(2) Unlike translation, a rotation acts around a point. Angle, angular velocity, angular acceleration and a moment are around a point. We shall take the point to be the mass center of the body (the geometric center of the basketball).

(3) With that build up, the mathematical definition of a moment is as follows:

> A moment is a force multiplied by a moment arm where the moment arm is the distance between the line of the force and the point about which you take the moment.

(4) The moment arm of the force is where the size and the shape of a body enter into the physics. Indeed, the translation of the body

is independent of size and shape whereas rotation depends on size and shape.

Imagine that your hand exerts a force on a basketball. You apply the force in some direction. If you were to extend a line along the direction of the force, calling it the line of force, that line may or may not pass through the center of the ball (again, see figure on Moment and Force). If the line of force were to pass through the center of the ball, the moment arm would be zero and you would not be applying a moment. In that case, you would say that the force is not producing a moment. On the other hand, if the line of the force were not to pass through the center of the ball, the perpendicular distance between that line of force and the center of the ball would create a moment arm. The force multiplied by the moment arm is equal to the moment that you are applying to the ball. In that case, you would say that the force is producing a moment. That moment, in absence of other influences, would impart an angular acceleration to the ball. In other words, the moment would increase the spin of the ball from whatever level of spin it had at that instant.

Finally, let us consider Newton's Law of Gravitation. Like the first three laws, we can extend it so that it applies not just to particles but also to bodies. That being the case, the two bodies that we are the most interested in here are the basketball and the earth. In particular, we refer to the force that the earth exerts on the ball as the **force of gravity** on the ball or just the ball's **weight**.

In summary, Newton's Laws describe the forces and the moments produced by the forces that propel a ball throughout the course of its motion. First, over the pre-launch stage, the body moves the ball in some appropriate way ending with launch conditions. Then, over the post-launch stage, the ball follows a ballistic (parabolic) trajectory like that of any other object under the force of gravity. Finally, the ball either does or does not enter the basket.

In the next section, we will tie this together. We will see, from a dynamics perspective what the best launch conditions are.

The Best Launch Conditions

As I said earlier, it comes down to a single instant of time, when the ball leaves your fingertips. The launch conditions dictate the fate of the ball's trajectory. Of course, there are many different possible sets of launch conditions but only one best set of launch conditions. By best, I mean the highest probability (likelihood of success) free throw. In this chapter, I put aside how to achieve the best launch conditions in any detail. Our concern right now is just to learn what the best launch conditions are. We have already found that that the launch conditions consist of launch positions, launch velocities, and launch angular velocities. Let us go through each of them, one-by-one, and see what the best ones are.

1) Launch position components (x, y, and z)

Say that you are standing right up at the free throw line. If you were standing in the middle of the line, x would equal to the distance between the center of the basket and the free throw line, and y would equal to zero. The z coordinate depends on from what height you launch the ball.

The horizontal distances x and y: The best launch positions are from the center of the free throw line where the distance of travel in the horizontal plane is minimal. Thus, $x = 13.75$ feet and $y = 0$ feet.

The vertical distance: The best launch height z is as high off the ground as you can achieve without losing any significant amount of consistency in your shot. The reason that the probability of success increases with the height of the launch has to do with the nature of the best shots with different launch heights. It turns out with different launch heights that the best ball trajectories still look extremely similar. All of them enter the basket at forward entry angles of about 45 degrees. For example, compare a launch height of 6 feet (z = -1.2 meters) with a launch height of 8 feet (z = -0.6 meters). When launching the ball at 8 feet, the ball is still under the player's control when its height is between 6 feet and 8 feet whereas, when launching the ball from 6 feet the ball is flying through the air during that time. When launching the ball at a lower height, you have to be more consistent to achieve the same free throw percentage (see figure of Optimal free throws at different launch hights).

1) Launch velocity components (v, α, and β)

2) The launch velocity components are the speed v, the forward angle α, and the side angle β.

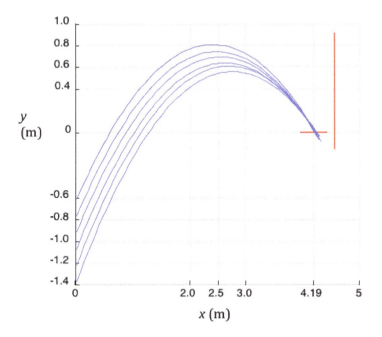

Optimal free throws at different launch heights

Speed: The best launch speed is the smallest. We also call this the softest shot. There are several reasons why the smallest speed ultimately produces the highest probability of success. Here are a two of the reasons.

- The smallest launch speed results in the smallest (softest) entry speed into the basket. With the softest entry speed, collisions of the ball with the rim and the backboard result in the smallest velocities after the collisions and the ball staying closer to the basket thereby increasing the likelihood of the shot being successful.

- The smallest launch speed also requires the least effort to produce on the part of the shooter. Shooters tend to be able to shoot with greater consistency as the effort they need to exert decreases. This is most evident when the player shoots the ball far from the basket. We speak of the player as having a "range," referring to his ability to control the ball at large distances from the basket.

Note that consistency does not necessarily decrease with increasing distance from the basket when the distance is small. Near the basket, the player's shot becomes "delicate." A few players have reportedly found the free throw to be "delicate" at the free throw line but this more often occurs much closer to the basket than at the free throw line.

Forward angle: The forward angle provides the trajectory of the shot with its arc and the best forward angle with its best arc. A too small forward angle produces a flat shot and a too large forward angle produces a high arch. There are two reasons why a too small forward angle and too large forward angle are problematic.

- The free throw that has the best arc also has the smallest launch speed. It requires the least effort to produce on the part of the player. Thus, the reasons for the best arc are the same here as for the best launch speed (explained above).

- The free throw with the best arc is also least sensitive to errors in the forward angle. This means if you shoot the ball with a forward angle that produces an arc that is in the neighborhood of the best arc, that the resulting error in the distance of travel of the ball is the smallest. This is easy to explain. Imagine two extremes. First, imagine shooting a flat shot. Your forward angle is small (see figure of Intensity of the Forward Angle). If you increase it by say 5 degrees, not changing anything else, the ball will travel farther. Instead of a swish shot, the ball may hit the back of the rim. Next, imagine shooting a high-arc shot. Your forward angle is large. If again you increase the forward angle by 5 degrees, not changing anything else, the ball will travel less far. Instead of a swish shot, the ball may hit the front of the rim. The point is that the sensitivity of distance of travel to forward angle goes from the distance of travel being too far to not far enough as you go from shooting a flat shot to shooting a high-arc shot. In between, when imparting a shot that has about the right amount of arc, the error of the distance of travel is the smallest. In other words, the free throw with the best arc is insensitive to errors in the forward angle. This means, you can achieve the same shooting percentage with the least consistency when launching the ball in the neighborhood of the perfect level of arc.

Side angle: The best side angle is zero. In other words, the best free throw is toward the center of the basket.

3) Launch angular velocity components (ω_x, ω_y, ω_z)

Axis of spin: In the best free throw, you are shooting the ball directly toward the basket, in the negative x direction. The free throw line is perpendicular to that. It lines up with the y-axis, the same axis about which you want the ball to spin.

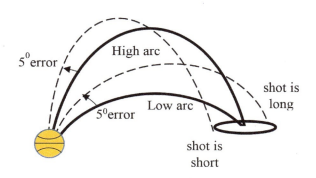

Intensity of the Forward Angle

Spin: About three revolutions per seconds of backspin is best. Incidentally, we refer to one revolution per second as a Hertz, abbreviated Hz, in other words 3 Hz of backspin is best. Let me explain why. Like with the softest launch speed, and with the best arc, the best level of backspin causes the ball, when entering the basket and colliding with the rim or the backboard, to stay the closest to the basket after the collision. To understand why this is the case, we need to understand better what happens when a spinning ball collides with a surface.

Imagine that you are still standing at the free throw line but that you bounce the ball on the floor two times (see figure of Ball Bounce). The first time you bounce the ball straight down with no spin. It comes back to you straight up. Next, you give the ball some backspin about the y-axis when bouncing the ball straight down. What happens? It bounces off the floor toward you instead of straight back up. The spin changes the way the ball bounces off the floor. Why is this?

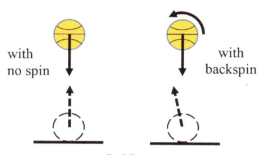

Ball Bounce

Now imagine that you happen to have with you a high-speed camera and that you use it to look at what happens to the ball over the tiny collision time of the ball with the floor. You place the camera on the floor beside you, facing it toward the ball, focusing in on the contact point. You will now see what is really occurring when you give the ball backspin. Just before the ball strikes the floor, the point on the ball closest to the floor is moving away from you (toward the hoop). Then, as soon as the ball strikes the floor, the floor reacts to try to prevent that point on the ball from sliding along the floor away from you (See figure of Ball Contact Over Milliseconds). It exerts an opposing contact force f on the ball that is toward you. The opposing force produces a moment around the y-axis of the ball that opposes the rotation of the ball (Law 2 for the Rotation of a Body). It causes the backspin of the ball to decrease. Even though this happens very quickly, before the end of that collision, the velocity of the contact

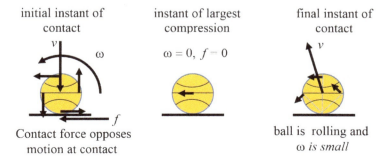

Ball Contact Over Milliseconds

point on the ball reaches a value of zero. The ball actually rolls off the floor. Since the force on the ball over that remarkably small amount of time was toward you, the ball moves toward you when it bounces back up (Law 2 for the Translation of a Body).

When shooting the free throw, this same change in behavior occurs as the ball strikes the hoop or the backboard. First, imagine that it strikes the top of the front part of the hoop. Just after striking the hoop, the ball with no backspin would be moving forward (away from you) and upward. If the ball were to have some backspin, its forward speed would decrease causing it to stay closer to the basket. Next, imagine that the ball strikes the backboard. Just after striking the backboard, the ball with no backspin would be moving toward you and downward. If the ball were to have some backspin, its downward speed would increase. Again, it would stay closer to the basket.

In summary, I have now explained to you what the best launch conditions are, going through them one-by-one. I did not give you any idea how to achieve them. However, you now know, at the very least, what to shoot for (excuse the pun). Let us now go over a few of the broad ideas that pertain to the dynamics of the shot during the pre-launch stage where the focus is on the best body motion.

The best body motion

From the dynamics perspective, the body is merely a set of articulated members connected by joints. The best body motion is about how to best articulate the different members of the body. This is a complicated problem, not just because of the number of joints and the number of possible articulated motions, but also because of the dependency of the articulation on the relative strengths of the joints, which plays out differently when the body is located at different distances from the basket. Let me explain.

1) Individuality

It started not long after you were born. You were on your back learning that you can move your arms and kick your legs. This was the beginning of a journey of building up body movements from simple to complex. You learned how to coordinate between these movements, wiring your neurons to the point that you presently enjoy. You learned how to reach, kick, stand, throw, walk, jump, etcetera. The points are as follows:

- The multiplicity of joint angles and members creates an extraordinarily large number of ways of achieving the same movements, such as those that make up a free throw. We refer to the combination of joint motions, one relative to the other, to achieve a particular sequence of movements as coordination.

- The coordination of each individual is unique. The body has wired the coordination between the joints to achieve the relative movements.

- We can change the wiring but it takes lots of practice to "reprogram" it. This is why we say that it is hard to change one's habits.

- However difficult it is to change one's habits, the change is different for each individual.

The overall point here is to recognize the individuality of the movements of the player and that this calls for different ways of modifying one's habits when seeking to improve one or more movements. However, it is even more complex than that because the way in which we throw the ball, for example, depends on the distance the player is from the basket.

2) Gaits

We most commonly refer to the relative motion of articulated joints as gaits. When a human moves, he starts with the walking gait, then the jogging gait, and finally the running gate. They depend on speed. When a horse moves, he starts with a walk, then a cantor, a trot, and finally a gallop. Again, they depend on speed. The gaits change because of the relative strengths of the joints that move the different members of the body.

For example, the joints that articulate the members of a horse differ significantly in relative strength, most notably between the hind legs and the front legs. As the horse needs to move quicker, it draws more from its hind legs than its front legs, changing its gait accordingly.

The human is very similar. The joints that articulate the members that make up its lower body are much stronger than the joints that articulate the members of its upper body. In the case of the human, it is a trade-off between strength and speed. Its upper-body members move and react quicker than its lower-body members react whereas its lower-body members are much stronger than its upper-body members are. In basketball, I have observed easily some five gaits as the shooter moves from near the basket to a half-court shot. Close to the basket, the shooter tends to draw on his upper body almost exclusively. He or she increasingly draws on the lower body as the distance to the basket increases. All of this is extremely relevant because it explains some of the broad themes in coaching that are pertinent to the individuality of the player:

- The relative strengths of the joints vary from player to player

- The relative strengths that dictate the best gait for an individual player depend on the distance from the basket.

Ultimately, any improvement in the coordination of a movement, such as a free throw, is a modification of a current practice. As such, it depends on the player's starting point, the way in which the player currently coordinates his or her body, which is unique to the player, learned over a lifetime, and therefore successfully changed by carefully identifying the changes that can make a difference. How one approaches this is critical.

Feedback

Again, imagine that you are standing at the free throw line shooting a free throw. How do you naturally approach the process of improving your shot? Do you focus your attention on the individual joints in your body or do you shoot the ball, see what happens, and adjust your shot accordingly. I suspect that your answer is that you do the latter but that you consider the former important, too. Indeed, seeking feedback is natural while seeking best practices can be helpful, too. In my view, Bob Fisher has written this book with the right balance between the two. While the former provides the quickest way to learn, the latter provides you with best practices that you can perfect with the greatest consistency. Furthermore, there is more feedback occurring in the free throw than you may first suspect.

Below I say a few words about the nature of feedback, from a dynamics perspective.

The movement of things—from robots to humans—results from applied forces. The body prescribes the forces to achieve desirable movements. The feedback is always present one way or another to ensure that the desired path be followed. The feedback can be part of the device, such as in a robot or a human, or it can be external to the device. A railroad track ensures that a train follows the right path. A pipe guides water from one location to another. Without feedback to guide the movement of an object, the object invariably goes off course. In short, nature does not generally apply forces with sufficient accuracy to guide a motion without feedback. Feedback, whether external or internal, is nearly always present in motion that follows a desirable path.

When engineering the control of the motion of a body, the engineer likes to say that a control force is composed of two parts: the tracking force and the regulation force. The tracking force is the force that moves the object from one point to another, at least in theory. It is the ideal force, if everything was perfect, that would take the object along its desired path. However, by itself, it does not work since the world is not perfect. The regulation force is the corrective action that requires feedback. The feedback comes from a measurement that provides an indication, one way or another, of the error present in the desired path. The engineer says that the "control system," is responsible for applying the force. It feeds back an error to the device to modify the force, to regulate it, so that the object can follow a path with satisfaction.

In the free throw, all of the feedback appears to occur during the pre-launch stage of the shot. Once you launch the ball, its destiny has been determined. Thus, the question arises how you acquire your feedback during the pre-launch stage. To answer this question, let us consider again the launch conditions and see how you acquire feedback for each one.

Start with the launch positions. The x coordinate is your distance to the center of the hoop. You acquire it before you take the shot by walking up to the free throw line. The y coordinate is your lateral distance from the center of the free throw line that you, again, acquire when walking up to the line. You determine the z coordinate, corresponding to the height of the launch of the ball, by your body motion during the shot. All three have "tracking" values, that is, ideal values. To achieve ideal values, you adjust them through feedback in some way. Where does the feedback come in? Next, consider the launch velocity components, specifically, speed, forward angle, and side angle. The ball starts at rest and then accelerates through a series of body movements until you launch the ball at a desired speed, forward angle, and side angle. All three have their desired values but how do you perceive and correct for errors in each? Finally, consider the launch angular velocities, specifically the ball's backspin about the y-axis. You want to impart about 3 Hz of backspin.

The question of feedback is an interesting one because there are several kinds. Furthermore, recall that the goal is ultimately to gain as much consistency as possible. Therefore, the kinds of feedback that you select result from their ability to facilitate a higher level of consistency in your shot. The four main kinds of feedback are:

- External markers
- Kinesthetic memory
- Body positioning
- After-shot assessment

External markers: These are geometric points that you focus on during a shot. It could be the free throw line, the front or back of the rim, or the top of the backboard, to name the most obvious ones. External markers provide feedback that can improve some of the launch conditions but not all of them. The launch conditions fall into two categories: geometric and kinematic. There are seven geometric launch conditions and two kinematic launch conditions. The geometric launch conditions are the three launch positions (the coordinates x, y, and z), two of the three launch velocities (the forward angle α and the side angle β), and the axis of the backspin (resulting in $\omega_x = \omega_z = 0$). The kinematic launch conditions are the speed v and the backspin ω_y. The external markers provide you with feedback for the geometric launch conditions but not as much for the kinematic launch conditions. The free throw line helps you with x. The front or back of the hoop helps you line up your shot, which helps you with y and the side angle β. External markers also help, to a small degree, with the in-plane geometric launch conditions, namely, with z, the forward angle α and with lining up your backspin axis (so that $\omega_x = \omega_z = 0$).

Kinesthetic memory: Kinesthetic memory is your memory of the motion of your body over the pre-launch stage. This provides you with the feedback that you need in order to impart the kinematic launch variables consistently, namely, the speed v and the backspin ω. Kinesthetic memory is also responsible for the ability to repeat a movement in time consistently. Most players shoot a ball in a two-step sequence, wherein they bring the ball to shoulder position after which they launch it. Others do this in one simultaneous movement. The kinematic memory required for each differ; the former is simpler and the latter is quicker.

Body positioning: Body positioning refers to the various positioning that you do during the pre-launch stage with little or no regard to how that might change over the course of the pre-launch stage. Examples of body positioning are the positioning of your feet and shoulders, keeping your elbow in, and the positioning of your hand and fingers on the ball. Body positioning helps with the geometric launch conditions and less so with the kinesthetic launch conditions.

After-shot assessment: The after-shot assessment typically refers to the assessment of external markers immediately after you shoot the ball. This is the most frequent kind of feedback used to improve a shot. For example, an external marker could be the ball itself. You could watch where the ball enters the basket. Does it enter the basket toward the front of the hoop or toward the back of the hoop? It turns out that the best shot enters the basket toward the back of the hoop. The reason is that the front of the hoop is less "forgiving" than the back of the hoop. An after-shot assessment can also provide feedback beyond that achieved by an external marker. Filming your shot provides you with the opportunity to assess your body positioning and the overall level of consistency in body motion that you are achieving.

Consistency of the launch conditions

I already mentioned something about the consistency of the launch conditions. Recall that I mentioned to you that the forward angle, when it is set to give you the best amount of arc, is insensitive to distance of travel. That was just one of the launch conditions. More generally, it is important to understand how consistent you need to be with each of the launch conditions.

While you need to impart all of the launch conditions with some minimal level of accuracy, some are more challenging or more important to impart accurately than others are. I discuss here the most important and challenging ones. In particular, I will focus on the launch velocities. Before discussing them, let me say a few words about the other launch conditions.

First, consider the launch positions. Although these are important, they tend to be easy to impart consistently. You set the x- and y-coordinates accurately by external markers and your body motion tends to end in a consistent z coordinate. Next, consider the launch angular velocities. These, too, tend to be set with sufficient consistency through your body motion. It is not that important to set the backspin of the ball to precisely 3 Hz nor is it particularly important for the backspin to be precisely about the y-axis. This leaves the launch velocities for consideration. These are the most important launch conditions to prescribe precisely and consistently. So, let us consider them in some detail.

Recall that the launch velocities consist of the speed, forward angle, and side angle. The question arises how they relate to the consistency of your body motion during the pre-launch stage. In particular, what is it about body motion that you can more easily perform consistently or that is more difficult to perform consistently?

Consistency is about repetition. It is a spatial and temporal (in time) sequencing between the different movements of the members and the joints in a body, one relative to the other. Depending on your heart rate, you may naturally perform the spatial sequencing more slowly or faster. You can think of body positioning as responsible for the spatial part of your consistency and kinesthetic memory with the temporal (time) part. Furthermore, the spatial sequencing of these motions occurs in a body frame. In other words, you can do this while pointing your body, overall, in any direction. The speed of the spatial sequencing and the direction in which the body is pointing when executing the body motion do not depend on body positioning. Your body records the speed of the spatial sequencing through kinesthetic memory and does not record the pointing of the body toward the basket in the sequencing. Both of these considerations have an impact on how you launch the ball with a consistent speed, forward angle, and side angle.

Launch speed: The spatial sequencing of the body is important to launch speed but more than that, so too is the speed at which you perform the spatial sequencing important. Indeed, consistent positioning of the body and good kinesthetic memory are both required to achieve a consistent launch speed.

Forward angle: We already discuss this, but can now add that it depends on the spatial sequencing of the body motion but not on kinesthetic memory.

Side angle: The side angle depends on spatial sequencing of the body as well as on the pointing of the body toward the basket.

In summary, the three major launch conditions depend on spatial sequencing. However, the launch speed depends on kinesthetic memory, as well, and the side angle depends on pointing the body, as well, making them particularly difficult to perfect. Indeed, launch speed and launch angle require special attention.

Part 3

Chapter 4
Objectives

*"The more expertise and experience people gain,
the more entrenched they become in a particular way of viewing the world."*

Adam Grant

Switching to a physics-based perspective caused me to reappraise my shooting objectives. Here are the essential elements I wanted:

- Biomechanically friendly

- Generate force as efficiently as possible

- Quick

- Provide feedback

- Provide maximum allowance for movement variability

Generating Force Efficiently

Growing up in the era of 15-foot jump shooters, I learned to bring the ball up to my forehead area and start the shot from there—basically, a one-two motion. Here is the sequence:

- Jump and bring the ball up to the forehead area

- At the peak of your jump—shoot

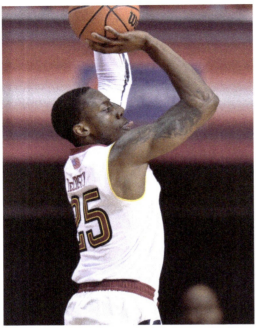
(Aspen Photo/ Shutterstock.com)

This was before the three-point line existed. Most shots taken from the perimeter were mid-range shots because there was no incentive to shoot farther out.

The study done in 1981 by Peter Brancazio, a physics professor at Brooklyn College, encouraged a high release. Brancazio determined that adding two feet to the height at which a shot leaves the player's fingers increases the success rate by 17 percent.[1] By jumping and releasing at the apex of your jump, you increase your odds of making the basket.

One-hour Record Demands Efficiency

When I first started shooting free throws, I used a high release. This changed when I started practicing to break the Guinness World Record for most free throws in an hour. That record was a challenge. I attempted it several times without success. Here is an email I sent to my friend Ryan Noel after one such attempt:

"Did poorly Sunday—definitely a learning experience. Learned the index finger release is not useful for volume shooting. I used it with both the left and right hand. After 10 minutes, my fingers started going numb. As you might suspect, this tends to make shooting rather difficult. With 20 minutes to go, I switched to the middle or middle and ring finger release (hard to tell when you can't feel your fingers). After a few minutes, the feeling returned to my hand. Finished with 1,664. Five minutes later I couldn't raise my right elbow above shoulder level without pain (it took a sports drink, three hours and two beers before I could). What happened?

With the index finger release, I was bringing my elbow under the center of the ball. Maybe this affected some nerves/muscles?"

Have you ever tried shooting when you can't feel your fingers? It's like wearing gloves. The numbness might have occurred because my hands were above my heart the entire time and my heart didn't pump enough blood up to them.

The one-hour record became a study in efficiency. How could I generate force as efficiently as possible? When fatigue sets in, it's hard to make shots. During training for this record, I lowered my release and became a one-motion shooter. It helped. Starting lower expended less effort and made my shot quicker. Win, win.

I eventually broke Perry Dissmore's record of 1,968; making 2,371 free throws in one hour. It is my most satisfying record because it was the most challenging. Over the course of an hour, I made a free throw every 1.51 seconds. What I learned during training for it was invaluable.

One Motion Versus Two

What did I learn? The old one-two motion of shooting is outdated except for mid-range jump shots. A case in point is Stephen Curry. Curry is a one-motion shooter, which is why he has such a quick release. The ball doesn't stop or pause at his forehead. Since he uses a smooth shooting motion, he generates force with a high degree of efficiency.

Many players use the one-two motion of bringing the ball to their foreheads and starting the shot from there. When shooting a three-point shot, their feet fly forward and their upper bodies recoil back because they are powering the shot primarily with their upper bodies. Players with a high release who slow the ball down at the forehead level are losing the early momentum of the ball moving upwards. With the ball's movement slowed, they have to use their upper bodies to apply more force. The result is a less efficient shot and typically, a flatter shot. The bottom line is the one-two motion shot is less efficient and slower.

The one-motion shot allows you to generate force efficiently and with less effort, which provides more control. As a result, your accuracy improves. Slow down to the extent the situation allows to enhance your precision further. Efficiently generating force will make your shot quicker while allowing you to feel as if you are shooting slower.

Biomechanically Friendly Wrist Motion

As I mentioned earlier, the wrist has muscles which allow it to move in any direction. These muscles have what is called "a natural pulling direction," which is when the movement occurs close to or in-line with that muscle. If the action is in a direction located between muscles, the nearest muscles create a tension against each other to produce the desired movement.[2]

My hypothesis is movements which follow the natural pulling directions of the wrist muscles provide a more consistent movement. Therefore, I want my wrist snap to be as natural as possible. In other words, no matter what position my body is in, I want a predictable wrist snap. Great shooters use a smooth, natural wrist snap and that is what I wanted.

The strongest position of the wrist is 10-15 degree ulnar deviation and 25-35 degree extension.[3] This location is easy to find. Just squeeze your hand and fingers together tight for a couple of seconds. Then relax your grip and open your fingers. Your wrist should be in its most athletic position. This position will provide maximum power to increase range and also help improve consistency.

The physics aspect of shooting involves two factors: distance and deviation. However, the biomechanics aspect of the wrist includes three elements: flexion/extension, radial/ulnar deviation, and supination/pronation.

Quick

One thing Stephen Curry has demonstrated is quickness rules. The speed of his release offsets his lack of size.

My reason for wanting a quick shot was rather straightforward. It allowed me to get up more shots. The more shots, the better my chance of breaking a record. Guinness doesn't care how many you shoot; they only care how many you make. Therefore, my shooting motion is game-like. How quick can I get a shot off? Pretty darn quick, even at my age.

Must Provide Feedback

Without feedback, how do you get better?

Feedback is essential. It helps to know what you did wrong. You must have feedback to learn what happened and how you can correct it.

Watching how the ball interacts with the rim and net will tell you what you did right or wrong.

Allowance for Movement Variability

As we've established, all shots do not need to look the same. Knowing that controlling the ball at release is the critical element allows you the various ways to accomplish this. You will find numerous ways to send the ball straight. With practice, you will learn to sense the weight of the ball and know where the center is. Controlling centerline makes it easier to throw it through the rim from any angle or position.

Push your limits. How can you learn what is possible if you don't stretch your boundaries? How differently can you shoot and still make the shot?

Conclusion

In basketball shooting, conventional wisdom has been to minimize variable movement. But is this necessary? When you utilize physics, it is not. If you want, you can try to make every shot look the same, but you don't have to.

We know for a fact when you control centerline, the ball goes straight without exception. The centerline concept is physics in action. Plus, we know there is more than one way to shoot and control the centerline of the ball.

Do you want to be a *pure shooter* or a *scorer*? What is the difference between the two? A pure shooter looks good shooting the ball; a scorer puts the ball in the hole. Which would you rather be?

(Petr Toman/Shutterstock.com)

Chapter 5
Be a Straight-shooter

"To do is to know; not to do is to guess."

Dr. Tom Amberry

While searching for common denominators which link different shooting methods, something clicked. I immediately texted Jessa Yaussi, a senior at Frankfort High School who understands my concepts. Smart young lady.

"I have a question for you. If you have the ball leave your thumb and little finger at the exact same time, do you regularly make it?"

"Yes," came the reply.

Me: "Good. How about leaving your index finger and little finger at the same time?"

A few minutes passed: *"Yes."*

Me: "What is the common denominator?"

Jessa: *"Funneling the ball?"*

Me: "Yes. But into what?"

Jessa: *"Centerline."*

Me: "Try this. Put your deep palm in the exact center of the ball before the wrist snap."

(What I didn't add is "the exact center of the ball *relative to the exact center of the basket.*" Jessa already knew this as she had heard it numerous times before.)

Jessa: *"OK!"*

Me: "That is the common denominator. The low in your palm is in the centerline of the ball and basket."

Jessa: *"Makes sense!"*

Understanding the "why"

You improve faster when you know what leads to success. Understanding why is a crucial component of getting better. If you don't understand the why, you tend to end up grasping at straws, experimenting endlessly and going down blind alleys, while becoming frustrated in the process.

A case in point: Zach and Curtis at *The Tonight Show with Jay Leno*. Zach and Curtis were brought in to rebound for Charles Barkley and me when we shot against one another. They arrived early and we did a run-through (with Chris, an associate producer, filling in for Charles). Afterward, I continued to practice and they rebounded for me. Taking a break, I encouraged them to shoot a few shots. Curtis went first and was erratic. I asked Curtis if he wanted a couple of tips and he readily said yes. I proceeded to explain his release and what needed to happen to make it work. He adjusted and immediately started knocking down shot after shot.

Zach's turn. Zach was releasing strictly off his index finger. I commented I had shot that way for years and it was somewhat inconsistent. "So, you're telling me, I'm screwed?" Zach asked. "Not at all," I replied. "That method works. It just wasn't the best for me. To make it work, you must keep your index finger going straight toward the basket at the moment of release. There are three ways you can keep your finger going straight." I went through them quickly and Zach resumed shooting. Within seconds, he was nailing shot after shot. The look Zach gave me was priceless. "Where were you when we were playing in college?"

Experience Helps

Curtis and Zach had played college ball. Both had put up thousands of shots. They had put in the practice and the time. The element they needed for improvement was knowledge.

Connie has asked me how long it takes me to turn someone into a great shooter. The answer is, it depends on the player's experience. Beginners take longer because they haven't put in the time necessary to become skilled enough to implement the knowledge consistently. Zach and Curtis were on the other end of the spectrum. They each had over a decade of training. With them, the lacking ingredient was knowledge.

KPT

Virtually every year an article will appear lamenting the woes of some player's free-throw shooting and communicate the idea that some players just can't shoot free throws.

This line of thinking is absurd.

Shooting is a skill. Since it is a skill, it is something you can improve. All it takes is KPT—knowledge, practice and time. The missing component for advanced players who struggle with shooting is knowledge. Increase that and you will have improvement.

Stagnant Shooting

In 2009, John Branch of the New York Times wrote the article *For Free Throws, 50 Years of Practice is No Help*. Branch points out the rate at which players make free throws has remained relatively unchanged for 50 years. In the article, according to Ray Stefani, a professor at California State University, Long Beach, improvement in sports depends on a combination of four factors: physiology, technology or innovation, coaching, and equipment. "There are not a lot of those four things that would help in free-throw shooting," Stefani said.[1]

What if he is wrong?

What if coaching could be improved? Is it possible shooting instruction initially missed the mark and we've been teaching a less than optimal approach for the past 50 years?

Before you answer this question, take a minute to watch the video on YouTube of "dad teaching kid to throw ball commercial." This video is a commercial for the Volkswagen Passat,[2] and it is hilarious. What makes this commercial funny is because nobody would teach their kid to throw a baseball like that. However, this is how we teach kids to shoot a basketball. Does this make any sense?

Old School

Fifty years ago, the most common release was the index-finger release. Bill Sharman, Hall of Fame player and coach, published *Sharman on Shooting* in 1965 and advocated the thumb, index and middle fingers should control the ball.[3] This period is when the "elbow under the ball" teaching became popular. George Lehmann, who played in the ABA and NBA from 1967 to 1974, introduced the BEEF acronym—for balance, elbow under the ball, eyes on the rim, follow-through—during the 1980s. Lehman compared shooting a basketball to throwing darts.[4] With his teaching, the elbow under the ball was the key to a straight shot. The BEEF analogy became a favorite teaching tool and some coaches still utilize it today.

With this release, the thumb and middle finger provide force to the ball and direct the center of the ball into the index finger. This method does work and works well. The advantage of this technique is alignment; the elbow is under the ball and in line with the rim. However, the disadvantage of this method is control and range.

The game has changed. Fifty years ago, there was no three-point line. Shooting a basketball in the same manner as throwing a dart made sense because players weren't attempting many long-range shots. Today they are. Now you need range and accuracy.

The Game Has Changed

Switching to the middle and ring finger release increased my range by approximately five feet. This result proved to be the case with players I have worked with as well. Players switching from the thumb, index and middle finger release to the middle and ring finger release can expect to see a substantial increase in their range.

Why? Because releasing off your middle and ring fingers (or your middle finger) generates more power due to a more natural wrist snap. Shooting with the index finger release is similar to throwing a dart. Another analogy would be throwing a jab in boxing. Both have the body angled with the toe, hip, shoulder, and elbow in alignment. However, a jab is not as powerful as a haymaker. You don't knock people out with jabs. You knock them out with a power punch.

When you shoot using the middle-finger release, your wrist movement more closely resembles the biomechanics of throwing a baseball or football. You are generating more power via your wrist snap, which results in more range than when you use the index-finger release.

"Throw It, Don't Shoot It"

Throwing darts and throwing a jab in boxing are learned movements, as is shooting with your elbow under the ball and angling your body toward the basket. They are the same primary motion. They lack power and take considerable time and practice to master.

On the other hand, shooting with the middle and ring fingers leaving the ball last is more like throwing a baseball or football. In particular, the wrist motion is the same. Keep in mind; the wrist action is the most significant contributor of force to the shot. Using a biomechanically-friendly wrist motion gives you more power, which results in more range.

The key takeaway here is there can be carry-over from one sport to another. You can shoot a basketball using the same concept as throwing a baseball. When basketball season arrives, you don't have to reinvent the wheel. Time spent playing baseball will help your basketball shooting. Both sports require you to throw a round ball accurately. One happens to be bigger than the other.

The adage, "Shoot it—don't throw it!" should be changed to "Throw it through the hoop!"

This idea might be heresy to conventional basketball wisdom. But players would improve faster and shoot better.

Blind Alley

I received a call from Kalib from Bryan, Texas. He was 32 years old and had started shooting free throws. Kalib said when he began practicing, he was making 60-65%. Looking to improve, he watched a YouTube video[5] and what they said made sense, so he switched to using the "Finger," a method in which the index finger is the last to leave the ball. Kalib immediately dropped to 50%, which didn't surprise him. He knew he might experience an adjustment period. However, two months later he was still shooting 50%. That's when he contacted me. We discussed the limitations of the index-finger release and I suggested he try either the middle or the middle and ring finger release.

The next day, Kalib emailed and said although it felt funny, he had made a higher percentage with both the middle and the middle & ring finger release.

What is the key takeaway from this story? Misinformation can be worse than no information. Learning to shoot can be like making your way through a maze. It is easy to get off the path which will lead you to proficiency.

(Aspen Photo/Shutterstock.com)   *(Photo Works/Shutterstock.com)*
The index-finger release may lack consistency for some players because it is more difficult to control the ball.

Starting on the Left

Many players are prone to missing left. To counter this tendency, some players start with the ball on the left side of their body. Kevin Durant does this somewhat, but Lonzo Ball does this big time. However, when you start on the left, you aren't going to be able to generate as much force. Starting the ball in front of your shooting shoulder allows you to add power by getting the weight of your body into the shot. The same way a shot-putter does.

A better way to counter the tendency to miss left is to apply force through the center of the ball at the moment of release. When the basketball hits the left side of the rim, it is because your control finger was on the right side of the ball at release. You must position your fingers to apply force in the middle of the ball as it leaves your fingertips. It doesn't matter if your arm and hand move inward while you shoot. This movement helps you generate power efficiently and situations may arise when more force is needed. What does matter is *you must control the center of the ball relative to the middle of the basket at the moment of release.* A majority of misses are to the left because most players start with their hand on the side of the ball and they don't get it centered by release. Therefore, the application of force was primarily on the right side of the ball, which results in a miss to the left.

Force Dictates Direction

Don't focus on mechanics. Do focus your attention on the ball. Specifically, controlling centerline of the ball and basket at release. Achieve this and you will send the ball straight every time.

Why? Because this is Newton's Second Law in action; the ball will go where you direct it. The ball goes straight when you apply force to the center of the ball in relationship to the middle of the basket.

Now the question becomes: How do you best control the exact center of the ball relative to the middle of the basket at the moment of release?

Your hand and fingers may not be the best tools for exerting force through the center of the ball, but it is the only tool we have. Your middle finger will distort if you are relying solely on it. It is best to use all your fingers, in the same manner as Kevin Durant and other great shooters do. The key is to control the ball. Great shooters apply the initial force with their palm and then use their thumb and fingers to guide the ball. They don't start with the ball on their fingertips.

Focusing on applying force through the center of the ball at release isn't traditional teaching. Standard instruction would have you focus on how you stand, where you have your elbow, putting your "hand in the cookie jar" or whatever else. Those items may have value and help improve your consistency. But you can make a shot while falling on your rear if you get the release right.

Straight Finger Techniques

However, moving your fingers at the basket while shooting is a great way to ensure a straight shot. Here are a variety of ways players accomplish this. Let's start with one of the most popular:

(1) Move your thumb forward

A favorite technique many great shooters use is to move the thumb forward while they shoot. Why? Because swinging the thumb forward rotates the ball, so their fingers are directly in the center of the ball at release. The thumb serves the same function as a rudder on a boat. A rudder steers the ship. Your thumb can guide the center of the ball into your fingers which leave the ball last.

(Richard Paul Kane/Shutterstock.com)
Note how swinging the thumb forward centers the ball in your fingers.

With the middle of your palm directly in the center of the ball, your lower thumb pad will be on the left side of the ball and the outside edge of your palm will be on the right. Apply force evenly to the ball.

Experiment and align your thumb with your index finger by the time the ball leaves your fingertips. Try pointing your thumb at the basket at release. How far ahead your thumb needs to be to center the ball will vary from player to player, but there is a position which is ideal for you. This location will not change—it is consistently in the same place in space and time. Find it and every shot will go straight.

I prefer to center my hand on the ball earlier rather than later. When I am a split-second late, I miss to the left. Which makes sense; when you apply force predominantly on the right side of the ball, it will go to the left.

(Aspen Photo: Shutterstock.com)   *(Aspen Photo: Shutterstock.com)*
Note the forward thumb position varies from player to player.

The thumb is the key. With your thumb forward, your fingers will move in a straight line toward the basket. With your fingers in the middle and moving toward the basket, you now have both the straight and centerline concepts working for you.

Moving your thumb forward encourages a natural throwing motion, which is more consistent. This physical action will eventually seem effortless. (This is the method of choice for Stephen Curry, James Harden, Klay Thompson and many others.)

(David Ortega Baglietto/Shutterstock.com)
Note the location of his thumb. This position causes his fingers to move toward the basket.

Summing Up

The keys to this technique are:

- Move your thumb forward to center the ball in your palm.

- Use an external focus, concentrate on the turn of the ball.

(Photo Works/Shutterstock.com.)
Note how the forward position of the thumb positions the other fingers in the center of the ball

- Turn the ball so your fingers are in centerline at the moment of release.

(2) Use the bucket-carrying angle

Our elbow has what is called a "carrying angle."[6] This 5 to 15-degree angle is evident when your palm faces forward and it allows us to carry objects without them banging into our legs. Therefore, with your hand facing away from you, your elbow will be out slightly. How much it will be out will vary from player to player.

Try this. With your hand in your line of sight, stretch your arm out in front of you with your palm facing away and your middle finger highest. Now bring your palm back closer to you, keeping your hand facing away and in your line of sight. You will notice your elbow went farther out. Try shooting from this position and you will find you will send the ball straight if you keep your hand facing the basket the entire time.

Why does this technique work?

Because having your elbow out rotates your palm and shifts the center of your palm slightly to the left and behind the ball. This change centers the middle of your hand in the center of the ball.

What is necessary with this technique is to forget your elbow and focus on the center of the ball. With your hand facing the rim as you shoot, you can use your entire hand to control the direction of the ball. Having your elbow out allows you to keep your hand facing the rim throughout the shot.

(Photo Works/Shutterstock.com)   *(Tumar/Shutterstock.com)*
Note the angle of the forearms.

However, remember to focus on what is essential—controlling and applying force through the exact center of the ball at the moment of release.

(Aspen Photo/Shutterstock.com)
Notice the angle of their forearms and how the center of the ball is directly over the middle of the shooter's palm.

(3) Turn your hand, so your palm faces the rim

Squaring the palm to the basket is another technique you can use to send the ball straight.

This technique is identical to the first technique; the difference is the thumb doesn't move forward as dramatically. It isn't a natural movement, or at least it wasn't for me. It is a rotation of the forearm as you shoot; using the radioulnar joint of your forearm to turn your hand toward the rim. This rotation of the forearm serves to center

(A. Ricardo/Shutterstock.com)   (Pavel Shchegolev/Shutterstock.com)
Note how the palm is square to (faces) the rim at release.

your hand in the centerline of the ball and basket. It can be learned and it works.

(Pavel Shchegolev/Shutterstock.com)   (Pavel Shchegolev/Shutterstock.com)
Note how these shooters are tilting their hands outward to center the ball in their palm before they start their shot.

Dr. Tom Amberry was the most accurate free-throw shooter in history and is in the Basketball Hall of Fame because of it. I remember watching Dr. Tom demonstrate his shooting motion and how quickly he rotated his palm to face forward as he extended his arm. "How do you get your hand turned like that?" I asked.

"You just do it," he answered.

(Dziurek/Shutterstock.com)
Note how this player has her hand directly facing the rim at the moment of release.

Turning your hand to face the basket is done by the radioulnar joint in your forearm. This joint is responsible for the pronation/supination of your hand. ("Pronation" is when your palm is facing away from you; "supination" is when your palm is facing you.) When you start with your hand on the side of the ball, it requires a minor pronation of your hand to get your palm facing the basket at release.

Turning your hand toward the basket can be a useful method of sending the ball straight. The key is to have your palm directly facing the rim and be applying force through the exact center of the basketball at the moment of release.

This method was somewhat unnatural for me at first. It took some time to master it. It is accurate and consistent when you get it right. To use an external focus, concentrate on applying force through the exact center of the ball.

(4) Fully extend your elbow

When you fully extend your elbow, it naturally squares your palm to the rim. Full extension of your arm squares your hand to the basket. But it does it in the last couple of inches. Try this now and you'll see what I mean.

This action may be why you hear some shooting instructors say, "Finish with your elbow above your eyes."

Moving your elbow above your eyes makes you fully extend your arm. Extending your arm squares your palm to the rim. Squaring your palm to the rim results in a straight shot.

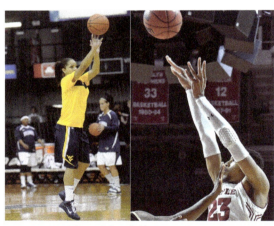

(Aspen Photo/ Shutterstock.com)
Note how the full extension of the arm faces the hand to the rim.

In shooting, there are many things you *can* do but *you don't have to*. In other words, certain mechanics may help you be more consistent, but instead of focusing on the mechanics learn what performing the mechanics accomplishes. Then you can focus on what is important. The important thing here is to square your palm to the rim before release. Whether you extend your arm or not is optional.

Watch Stephen Curry and you will see him shoot without fully extending his arm. In fact, there are times he flat-out short-arms it. But he does get his hand facing the rim before the ball leaves his fingertips.

A word of caution about fully extending your arm: It is possible to hyperextend your elbow doing this. To avoid that, rotate your arm slightly

(Aspen Photo/Shutterstock.com)
Note the partial extension of this player's arm. Full extension of the elbow isn't a requirement. Controlling the centerline of the ball at the moment of release is.

as you reach, so the tip of your elbow is pointed to the side, in the same manner as throwing a haymaker when boxing. This action will keep you from hyperextending your elbow.

(5) Start your shot with the ball in front of your shooting shoulder

With the ball in front of your shooting shoulder, your palm naturally faces the rim. To demonstrate this—bring your hand up to the middle of your chest, with your elbow by your side. Notice how the palm of your hand is at an angle to the rim. Now without moving your elbow, move your hand in front with your shooting shoulder. With your hand in front of your shoulder, your palm faces forward.

Larry Bird shot this way.[7] He started his shot farther out than most. He started off to the side of his head, which placed his hand more in-line with his shooting shoulder. This position allowed him to shoot naturally. What impressed me was the ease with which he shot. Bird made it look effortless.

(Richard Paul Kane/ Shutterstock.com)

(Richard Paul Kane/ Shutterstock.com)

Shooting in front of your shoulder has the advantage of alignment with your shoulder joint, which contributes force to your shot. Plus, by merely extending your arm, your hand will go straight—all you have to do is reach for the basket. The disadvantage is it places the ball and your hand farther away from your line of sight.

I should mention Bird could (and did) shoot with the ball closer to his line of sight. Variation is a commonality among great scorers; they develop the ability to shoot efficiently in an assortment of ways.

(6) Dart-thrower's wrist motion

This technique covers any shot where your fingers are moving at the basket, but your palm is at an angle to the basket. The tip-off to noticing this method is the palm is not facing the basket at the moment of release. This technique uses your wrist to control direction, somewhat similar to throwing a dart. Snap your wrist at the basket, so your fingers go directly toward the basket.

All the methods described involve snapping your wrist at the basket. What sets the dart-thrower's motion apart is there is no call for a particular hand placement on the ball. In other words, the fingers aren't required to be in the exact center of the ball at release. In theory, it is a simple technique. I included it here because it does work. However, it lacks control when compared to other methods. You rarely see this approach used today.

Summary

The common denominator to these methods is your fingers should be going straight toward the basket at the moment of release. This action is the most critical aspect of the straight-line concept. The above options are different ways you can make that happen.

Every shot doesn't have to look the same. Everyone has movement variability—we aren't robotic by nature.[8] We cannot replicate the same exact motion time after time. We can reduce movement variability with practice, but minor amounts still exist.

The reason for listing these different ways to send the ball straight is to permit you to experiment. You have options. You have choices. There is any number of things you *can* do but don't *have* to. You can learn to shoot numerous ways accurately.

All these techniques will help you send the ball straight. They do have something in common. We will cover this common denominator in the next chapter.

Chapter 6
The ANSWER

What is the common denominator of all great shooters? What is the one thing they do, no matter what release or method they use?

This question haunted me for years. After reading Fontanella's book, I started searching for the answer. There were times I doubted I ever would find it if there were one. There is such variation among players in the way they shoot.

Different Strokes for Different Folks

Watch any team warm up and you will see at least two or three different shooting styles among players on the same squad. This variation is what makes shooting interesting. Different methods work. Players do not have to all shoot the same.

This question emerges—is there a common link between the various shooting methods? And, if so, what is it?

Over the years, while at the gym, I have had many insights. Which is why I regularly carry a notebook—whenever I learned something new I would record it. Notes and diagrams ensured I wouldn't forget them. Plus, the act of writing further embeds the material in your brain. This tactic has proven to be helpful and I would encourage you to do the same.

My best insight answers the question of what shooting methods have in common.

Common Denominator

The most important aspect of shooting is this: At the moment of release, you must have *the low of your palm* in line with *your fingertip* which leaves the ball last and *the center of the rim*.

If you have these three items in the same vertical plane at release, you will send the ball straight because this is the straight-line thrust concept in action.

However, if you add one more item to this list, you will never miss left or right again. That item would be *the exact center of the ball*.

Sending the Ball Straight Every Time

When you have the low of your palm, your control fingertip, the center of the ball and the middle of the basket in the same vertical plane at release, the ball will go straight every time. This concept is physics in action.

Initially, this was hard for me to believe and trust to be true. However, time and experience have convinced me because it continually works without exception.

(Aspen Photo/Shutterstock.com)
Note the low point of his palm is in line with the exact center of the ball as this player is snapping his wrist. This alignment is the key to a straight shot.

When you drop the ball, it falls to the ground. Why? Gravity. The gravitational pull of the Earth pulls the ball downward. In the same manner, controlling the center of the ball relative to the middle of the basket is based on science. The ball is going to travel in the direction force is applied. This fundamental concept simplifies and quantifies shooting to its key ingredient. Control the center of the ball and you will send the ball straight without exception.

This knowledge simplifies shooting, but you still have to do it. Since there are multiple ways to accomplish this, it raises the question: What is the best way for you to control the center of the ball relative to the middle of the basket at release?

Get in the Plane

Try this. Before you snap your wrist, align the center of your hand in the same vertical plane as the exact center of the ball and the exact center of the rim. Do this and your fingertips will *move to the exact center* of the ball and basket at release.

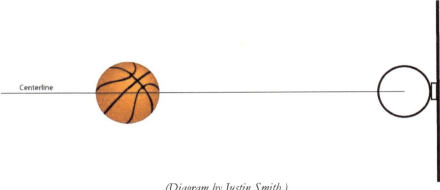

(Diagram by Justin Smith.)

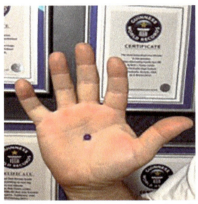

Positioning the middle of your palm on the centerline of the ball and basket before you snap your wrist will produce a straight shot every time.

How do you position the low of your palm in centerline?

Here's how. Face the basket. Using only one hand, balance the ball. Don't have it up on your fingertips but instead let the ball make full contact with the palm of your hand. Allow the ball to sit on the "cup" of your palm. Balance the ball in this area of your palm. The low of your hand should be in the exact center of the ball. Now shoot.

You will notice the ball goes straight each time. When it doesn't, the center of your hand was not in centerline at the moment of release. If you miss, it is because the low of your palm was not in the same vertical plane as the centerline of the ball and basket. The key to shooting is to balance the ball over the low point of your palm before you snap your wrist.

(LeVanteMedia/ Shutterstock.com)
Note how the low point of his palm is in centerline.

(Aspen Photo/Shutterstock.com)
Note how this player is balancing the ball in one hand. This positions the low of his palm in the exact center of the ball as he starts the wrist-snap phase of his shot.

Simple but Effective

This insight simplifies shooting. Your objective is to control the center of the ball. Balance the ball in your hand and then your fingers throughout the shot and the ball will go straight without exception. In other words, maintaining a balanced ball will result in a straight shot. Therefore, it is a matter of learning to balance the ball throughout the entire shooting motion, especially at the moment of release. What makes this difficult is the wrist-snap phase of the shot is happening right before the moment of release. It is easy to have the center of the ball shift during this time, which causes the ball to squirt. Learn to press against the exact center of the ball at the moment of release and you will be golden.

When you shoot, focus on balancing the ball in one hand. When you dribble, concentrate on controlling the ball with one hand. Anything you can do which will enhance your ability to balance and direct the center of the ball is going to be beneficial to you.

Apply Force through the Center of the Ball

By starting the shot with the low of your deep palm in the center of the ball, you will be applying force through the exact center of the ball at the moment of release. This concept results in the ball always going straight. However, it is *not a requirement* you start with the low of your hand in the center of the ball. You can still make shots without doing so and many players shoot without having their fingers or palm in the middle of the ball. But when they do, they are utilizing the Straight-line Thrust concept exclusively. Remember, with this method the fingers must move *in line* with or *parallel to* the pivot point of the wrist for the ball to go straight.

However, to use both the Straight-line Thrust and Centerline concepts, you must add the center of the ball into the mix. Doing both requires the low of your palm, fingertips, middle of the ball and center of the basket to be in alignment at release. Balancing the ball over the low point of your palm before the wrist snap ensures this will be the alignment at release.

Making Every Shot

To make every shot, we must account for distance as well. To do this, we have to add one more thing: The low point in your deep palm should be in the exact center of the ball relative to the apex of the shot and aligned with the middle of the basket. Squaring your palm to the top of the arc (or slightly higher) will guarantee the appropriate arc—assuming you apply the correct amount of force.

Try this: Visualize the ideal arc necessary for your distance from the rim. At a minimum, face your palm at a 90-degree angle to slightly above the apex of that arch. You can cock your wrist back farther, but at some point during the shooting motion, your palm should face somewhat above the top of that arc. Snap your wrist and you should have the launch angle right. Apply the correct amount of force and you are money.

Also, be aware that balancing the ball with one hand only will cover this facet as well because your palm will need to be facing the apex of the arc or even higher to control the ball adeptly. Balancing the ball in one hand takes care of the distance aspect automatically.

Physics In Action

The critical point here is the exact middle of your hand should be in the exact center of the ball before you start your wrist snap because this is a fundamental law of physics in action. If you are square to the hoop and align the low of your palm with the centerline of the ball and rim be-

(Aspen Photo/Shutterstock.com)
Note how this player is balancing the ball in one hand. This positions the low of his palm in the exact center of the ball as he starts the wrist-snap phase of his shot.

fore you start your wrist snap, the ball will consistently go straight, due to your last moving joint.

(Francesc Juan / Shutterstock.com)
Note how this player had the low of his palm and his fingertips aligned with the center of the ball and the middle of the basket at the moment of release. This shot demonstrates the direction your fingers move toward centerline may vary.

The Last Major Hinge Joint Dictates Direction

Since the release is the most important aspect of shooting, your last moving joint is going to determine the final direction of the shot.

Think of it this way. Let's say we are assigned the task of building a launching device to catapult basketballs into a basket. If this launching device had a minimum of three joints which apply force to the ball, which joint would impact the final direction of the flight of the ball? The last one—the one closest to the moment of release.

(Jamie Lamor Thompson/Shutterstock.com) Note the bent knuckles of Anthony Davis.

In theory, the joints of your fingers would be the last joints to affect the direction of the flight of the ball, but most players don't bend their finger joints when they shoot. However, the first and largest knuckle, the junction between the hand and the fingers—is commonly used to apply force to the ball by many players when they shoot. This knuckle area is in the same vicinity as the low of your palm. Your knuckles are the last major joint before the moment of release. Therefore, they will impact the final direction of the ball.

Since the release is the most important aspect of shooting, your last moving joint is going to determine the final direction of the shot.

Working Backward

We know the release is the most important aspect of shooting. Since this is the case, then it follows that the most important joint is the last one which affects the final direction of the ball before release.

The last major joint which affects the direction of the shot is your knuckles. The low of your palm is where your hand bends when you use your knuckles to bring your fingers forward. Positioning the middle of your palm in the center of the ball results in a straight shot because your knuckle is the joint dictating the final direction of the ball.

What if you don't bend the knuckles of your hand when you shoot?

Then the last major joint which would affect the final direction of the ball becomes the pivot point of your wrist. But the same concept applies. Align the pivot point of your wrist, with the centerline of the ball and basket before you snap your wrist and you will always be a straight shooter.

Not the Wrist or the Elbow

For years, I thought the pivot point of the wrist was the critical joint which dictated direction. At times, the pivot point of my wrist was in alignment with my fingertips, the center of the ball, and the middle of the basket, but occasionally it wasn't.

With my middle finger knuckle centered the ball invariably goes straight. For me, this is the key. Placing my knuckle in the exact center of the ball before I snap my wrist sends the ball straight no matter how I follow-through.

This insight sounds simple, but it works. If there is a secret to shooting, it is this: *Align the low of your palm with the centerline of the ball and basket before you snap your wrist.*

This concept works because no matter how you shoot, your fingers are going to cross the centerline of the ball and basket. Since the ball is a symmetrical sphere, your fingertips should be in the exact center of the ball when it leaves. This technique's outcome is you will be applying force through the exact center of the ball as the ball takes off from your fingertips.

Balance the Ball in Your Shooting Hand

As I mentioned earlier, a simple way to accomplish this is to balance the ball in your hand when you shoot. When you do this, you are combining simple physics with biomechanics and using them to your advantage. In essence, you are setting yourself up for success.

With your shoulders square to the basket and the ball balanced in the center of your palm; extend your arm, snap your wrist and your fingertips will be on the centerline of the ball and basket at the moment of release. Once your fingertips reach the center of the ball, they will lose contact because you have reached the highest point of the curvature of the ball. What direction your fingers approach the middle of the ball makes no difference. All paths lead to center of the ball.

*(Aspen Photo/Shutterstock.com)
Note how she is controlling the ball by gripping it with her little finger and thumb. Though unorthodox, this centers the ball in her palm, which will likely result in a made shot.*

Chasing Centerline

Chasing centerline is a term I use to describe the concept of the last joint moving the fingertips to the middle of the ball at the moment of release. Here is an exercise which will demonstrate this idea. (For clarity, I will be using the middle finger as the release of choice in the following instruction.)

Square up to the rim. As you start your shot, your hand can be in any position on the ball. Since we are utilizing the centerline concept, your initial hand position makes no difference. What is important is to have your full hand in contact with the ball to gain complete control. Visualize the point of the ball which would be the exact center of the ball. As you start your shooting motion, apply force toward this spot with your fingers (the exact center of the ball).

(Diagrams by Justin Smith)

As the ball moves up and away from you, continue to apply force toward this imaginary point which represents the exact center of the ball. Continue this to release. At release, your middle finger will be on centerline of the ball and basket, which will result in a straight shot.

Feedback

Misses occur when your last moving joint and your control finger are off-line of the centerline of the ball and basket *at the moment of release*. A miss to the left will happen when your knuckle (or last moving joint) is on the right side of centerline at the moment of release. Vice versa, a miss to the right will occur when the middle of your hand is on the left side of centerline at the moment of release.

Feedback for the Runner

When shooting a 10-15 foot shot while moving toward the basket (commonly called a *runner* or *floater*), your hand should start on the side of the ball. This position enhances your ability to control momentum. Visualize the point which is the exact middle of the ball and apply force toward it as you shoot. Your fingers will be moving from the side to under the ball at the moment of release.

Feedback for this shot is different. With your fingers moving to centerline from the side, a miss to the left will occur when your middle finger is behind the center of the ball at release. A miss to the right will result when your middle finger is in front of the center of the ball as it leaves your fingertips.

Expanding Your Options

It bears repeating that the direction that your fingers move towards center makes no difference. Since the ball is round, all paths lead to center. This allows you tremendous movement variability and will enhance your ability to make shots which will amaze the crowd. When you master this concept, you will be virtually unguardable. This will transform your shooting like nothing else.

When you start your wrist snap with the middle of your hand in the exact center of the ball and have your fingertips cross centerline at release, you can be moving *laterally* and the ball will go straight. However, this does involve the additional element of momentum which must be accounted for, but you can learn that. We will cover shooting while moving in a later chapter.

Wrapping Up

What is the common denominator of great shooters?

Controlling the center of the ball is the crucial element of shooting a basketball. Most excellent shooters position the middle of their palm on the centerline of the ball and basket before they execute their wrist snap. Placing the center of your palm on the centerline of the ball and rim before you snap your wrist will produce a straight shot every time. A miss to the left will occur when the middle of your palm is to the right of the centerline as your snap your wrist. Vice versa, a miss to the right happens when the center of your hand is to the left of centerline at release.

(Aspen Photo/Shutterstock.com)
Note how this player is controlling the ball with three fingers, with his middle finger in the center of the ball.

How do they accomplish this? By balancing the ball in their shooting hand. Missed shots are caused by not keeping the ball perfectly balanced during the shooting motion. Balancing and controlling the centerline of the ball throughout the shot results in a straight shot every time.

Try it. It works. This insight is the answer you have been seeking. If there is a secret to shooting, this is it. Use it to your advantage. Develop this technique and you will be amazed at your increased ability to make shots.

Chapter 7
Funneling

"Innovation rarely comes from the established institution. It's always a graduate student, or a crazy person or somebody with a great vision."

Eric Schmidt, Google Chairman

Funneling is a technique used by the best shooters in the game. What is it?

Funneling is positioning your fingers and thumb to channel the center of the ball into your control finger. Similar to using a funnel to focus liquid through a narrow opening.

Funneling a basketball is one way to use physics to help you make shots. We have established the fact that applying force in the exact center of the ball at the moment of release will result in a straight shot every time. Funneling is a technique which continually centers your control finger in the exact center of the ball at the moment of release. Funneling involves using your hand to cradle the ball so the last finger leaving the ball is in the exact center of the ball at release.

With funneling—as with virtually everything involved in shooting a basketball—you have options:

- You can use your thumb and middle finger to funnel the ball to your index finger

- You can use your thumb and ring finger to funnel the ball to your index and middle fingers

- You can use your index and ring fingers to funnel the ball to your middle finger

- You can use your index and little fingers to funnel the ball to your middle and ring fingers

They all work. You can master them all or strive to perfect one.

(Aspen Photo/Shutterstock.com)
Note how this player is using his ring and index fingers to funnel the center of the ball into his middle finger.

(CP DC Press/Shutterstock.com)   (Pavel Shchegolev/Shutterstock.com)
Note how the shooter on the left is funneling the center of the ball into her middle finger. The player on the right is funneling the center of the ball to his index and middle fingers.

When you practice, if you pay attention and focus on funneling, it will come naturally in a game. Learn to catch the ball with your hand positioned so you can funnel the ball to your preferred release.

(David Ortega Balietto/Shutterstock.com)
Note how this player's thumb points forward. He used his thumb to help funnel the ball to his index, middle and ring fingers.

Many people believe in using one method, but if you study consistently high scorers, you will notice they have developed the ability to use more than one release in game situations. They can react to the situation and still put the ball in the hole. Learning more than one method isn't difficult. Brynlee, my 7-year-old granddaughter, can shoot two different ways quite well. Both methods work. She thought she was shooting the same way each time until I pointed out the difference. She didn't care. She was just happy to see it go in the basket.

Here is what makes each method work.

Starting with the most popular, which is using your index and ring fingers to funnel the ball to your middle finger. The middle finger leaving the ball last is the most common release used by today's NBA players. Stephen Curry, James Harden, Kyrie Irving, Lebron James, Klay Thompson, Kawhi Leonard and Russell Westbrook are some of the best shooters in the NBA who utilize this technique.

One of the reasons this method is so popular is it provides excellent control, which makes it the technique of choice for three-point shooters.

(Aspen Photo/Shutterstock.com)
Note the configuration of this player's hand at release. This hand position funnels the ball into his middle finger.

Finger-leaving Sequence

Gripping the ball with your thumb and little finger provides exceptional control over the ball. The action of the thumb moving forward steers the center of the ball to your three middle fingers. Once your thumb and little finger leave the ball, the ring and index finger continue to control and funnel the ball into your middle finger.

It happens in a split-second, but it is a three-stage process. First, the thumb moves forward to help the little finger funnel the ball to the index and ring fingers. This positions your thumb to leave the ball at approximately the same time as the little finger. Next, the ring and index finger funnel the ball into the middle finger. Your middle finger should now be in the exact center of the ball and it leaves the ball last.

(Photo Works/Shutterstock.com)   *(Photo Works/Shutterstock.com)*

(Photo Works/Shutterstock.com)   *(Photo Works/Shutterstock.com)*

The stages of the shot: the ball leaves the thumb and little finger; the ball leaves the index and ring fingers; the middle finger leaves the ball last. This sequence happens in a split second.

As you look at pictures of this release you will notice the index, middle and ring fingers form the shape of a turkey's foot.

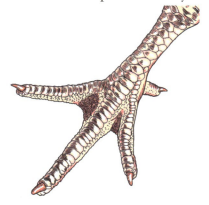

(Illustration of turkey's foot: Shutterstock.com)

Aspen Photo/Shutterstock.com)
Note how this player's middle finger is in the center of the ball.

Fingertip Touch

Is it possible to feel your fingertips leave the ball?

Yes, but it requires your full attention. I have found it is helpful to shut your eyes and shoot because this heightens your sense of touch. It is not easy and don't get discouraged if it seems impossible at first. It is possible; give it time. Over time you will develop the sensitivity in your fingers. Be persistent. You will get it if you stay with it.

Players utilize another release where the index and little finger funnel the ball to the middle and ring fingers. With this technique, the index and little fingers apply force and leave the ball at the same time. They push evenly against the circumference of the ball and funnel it to the middle and ring fingers.

Another release sometimes used is the thumb and ring finger to funnel the ball to the index and middle fingers. With this technique, the thumb and ring finger push evenly against the ball and guide the center of the ball to the index and middle fingers. A split-second later the index and middle fingers leave the ball last.

Funneling is a useful technique to utilize physics to help you make shots. The application of force in the exact center of the ball is an efficient way to be a straight shooter.

(Pavel Shchegolev/Shutterstock.com)
Note how this player is releasing the ball off his index and middle fingers.

(Pavel Shchegolev/Shutterstock.com)
Note how this player is using his thumb and middle finger to funnel the ball to his index finger.

Chapter 8

Long and Short

"You have to put in the time and do the research and then the connections are made and the revelations will come."

Jordan Ciambrone

During clinics, I tell players the distance aspect of shooting is easy; which it is in comparison to the deviation aspect.

Distance involves only two factors: launch angle and launch speed. The speed and angle at release create the arc of the ball. These two factors must be in sync with each other, depending on your location on the court. Your distance from the basket will determine what launch angle you need and how much force you should apply to achieve the proper arc.

Finger Position at Release

Where your fingers are on the ball at the moment of release will determine the angle of your arc. For example, for a two-foot shot, your fingers should be on the underside of the ball pushing upwards (at about a 72-degree angle)[1]. For a full-court shot, your fingers should be behind the ball if you are to have any chance of getting the ball to the rim.

Watching how the ball interacts with the rim and net provides feedback to what you did right or wrong. For example, if you back-iron the shot and the ball bounces out, what went wrong? The arc on your shot was too flat. Why? Because at the moment of release, your fingers were higher than they should be on the backside of the ball. How do you make the necessary correction? If you are shooting from the same distance again; position your fingers a little lower on the backside of the ball at the moment of release. Now the ball will travel upwards more than on the previous shot.

When you shoot an airball, your fingers were too far forward on the underside of the ball at the moment of release—unless your launch speed simply wasn't sufficient to get the ball to the basket.

When you miss long or short, adjust one of those two factors (launch angle or launch speed) to correct the problem.

Shooter's Touch or Larger Target?

Should you go for the shooter's touch or for shooting at a more prominent target?

In Fontanella's book, he ran through the equations which he used to calculate the ideal launch angle for the slowest moving ball when it nears the rim.

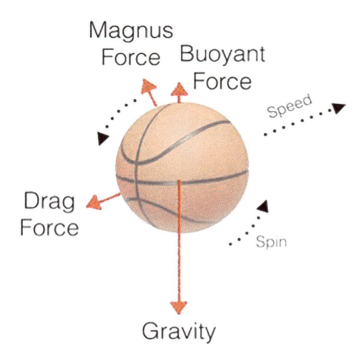

(Diagram courtesy of Aatish Bhatia.[2])

Fontanella used the four factors which affect the ball: gravity, drag force, Magnus force and buoyancy of the ball. He calculated a 6-foot player shooting a free throw should use a launch angle of 51 degrees and a 7-foot player should release with a launch angle of 48.7 degrees.[1]

A slow-moving ball as it nears the basket will give you a "shooter's touch" because any interaction with the rim will cause it to bounce softly and possibly fall through.

James Harden is 6 feet 5 inches and is shooting at 49.6 degrees,[3] which is spot-on for his height for the slowest moving ball as it nears the rim. Hardin has a shooter's touch because he is releasing with the optimum angle to achieve a soft shot. However, Stephen Curry does not. For a player of his height, Curry releases with a higher launch angle than average. According to the data from the NBA's SportVU motion tracking, Curry is shooting at 58.1 degrees.[3]

Curry isn't going for a slow-moving ball. He is opting for a more substantial target. With a higher launch angle, the area of the basket available for the ball to pass through increases. Opting for a higher launch angle, Curry is increasing his target size.

Many of today's NBA players are opting to shoot at a larger target. In other words, they are shooting with slightly more arc than necessary, thereby increasing the area size of the goal for the ball to pass through, rather than opting for a slow-moving ball as it nears the rim.

Best Left to the Individual

What this shows is shooting is an optimization problem and optimal throwing conditions are individualistic.

A study by Irina Barzykina confirms this. This study released Feb. 24, 2017, is called *The Physics of an Optimal Basketball Free Throw*. In it, the author states, "Some of the most successful players bear completely different conditions to the average player, which also confirms the fact all players have different consistency in release angles and velocities. Some might need more space for error in velocity and thus need a higher throwing angle, while others might aim lower because their velocity control is much stronger."[3]

Well said. In other words, shooting is an individualistic endeavor. We have differences which can be used to our advantage. For best results, you should utilize your strengths and find what works best for you.

But this is something smart basketball people already know.

(Viacheslav Nikolaenko/Shutterstock.com.)
Note how on a direct shot the ball enters the back half of the rim.

Danny Ainge, general manager and president of basketball operations for the Boston Celtics of the National Basketball Association, put it this way: "Everybody has different mechanics and sometimes you don't get as detailed because the great coaches understand the uniqueness of each individual. There is not one way that everybody shoots. Everybody has a bit of a different body, shoulder rotation, or getting your elbow, wrist, and follow-through under the ball. Everybody's bodies are unique. Even hand position, some guys can't spread their thumbs wide."[4]

One thing you shouldn't do is shoot with a low arc. A flatter shot results in a faster moving ball at the rim and more importantly, reduces your target size. Anything less than the arc required for a shooter's touch is a losing proposition because you have less area of the rim for the ball to pass through. It is better to have too much arc than not enough.

Experiment to Find What Is Best for You

Reading between the lines, what Barzykina's study suggests is we shouldn't be afraid to experiment. Otherwise, how will you find what works best for you? When you do discover something which works for you, if it doesn't quite jibe with what others are doing, don't think you have to change to "look good." Looking good and missing isn't as acceptable as making shots. Remember, "We are not selling jeans here."[5]

Although the basket is 18 inches in diameter, the angle the ball approaches the basket determines the size of the area the ball has to pass through. If you stood on a ladder and threw the ball directly down through the exact center of the rim, there would be 4.25 inches of clearance in all directions. Do the same thing with a women's ball and it would have 4.4 inches of clearance because a woman's ball is 9.2 inches in diameter.

(Diagram by Justin Smith)

Lower Arc Equals Smaller Target

While shooting a men's ball, when the ball comes in at a 55-degree angle the clearance is reduced to 2.5 inches.[6] Lower the approach angle to 45 degrees and the distance from the rim drops to 1.5 inches. At an angle of 33.3 degrees, the ball will not be able to pass through the basket without touching iron; there is zero margin for error.

Looking at these figures, it becomes apparent why shooting is challenging, which makes the prowess of today's top shooters even more impressive.

Using a higher arc corresponds to shooting at a larger target. However, the drawback is a higher arc requires the application of more force. This Catch-22 is yet another example of why shooting is an optimization problem.

Most Efficient Shot

"Consider the amount of force needed to launch the shot," Peter Brancazio wrote in the 1984 *Sport Science: Physical Laws and Optimum Performance*. "It is to the shooter's advantage to use as little force as possible."[7]

Brancazio was ahead of his time. A recent study

(Muszy/Shutterstock.com)
From this angle, the rim presents a larger target.

revealed throwing softer leads to an improvement in accuracy. In April 2017, this study appeared in Royal Society Open Science. Two professors, Madhusudhan Venkadesan of Yale and Lakshminarayanan Mahadevan of Harvard, used mathematical modeling to determine there is, indeed, a speed/accuracy trade-off. "What our work adds is a physics basis for why there is a speed/accuracy contrast," Venkadesan is quoted as saying.[8]

The Speed/Accuracy Trade-off

Leisurely shoot ten free throws. Then put ten seconds on the clock and fire-up ten free throws. You will quickly discover the faster you go, the easier it is to miss.

A reduction in accuracy due to an increase in speed is called the speed/accuracy trade-off. What I learned from speed shooting correlates to what Brancazio said: "It is to the shooter's advantage to use as little force as possible."

Using a least-effort shot makes it seem as if I am going slower. The more effort you exert, the more difficult it is to maintain your sense of touch. The slower you go, the more you can sense the ball leave your fingertips.

It is gratifying to have science verify something we have learned to be true. It adds credibility and removes doubt, especially when teaching others.

Summing Up

What is the bottom line here?

When you shoot, generate force as effectively and efficiently as possible. A proficient generation of power will provide maximum control, thereby improving your accuracy. Shooting in rhythm will increase your percentage. Rushed shots are something you want to avoid.

Should you shoot with more arc, so you have a more significant target, or should you try for a shooter's touch? This dichotomy is for you to figure out. What works for you may be different than the next person. You may have a gift for using the optimum arc for a shooter's touch, or you may have better results shooting higher so that you have a larger target. Find what you do best and know if it is a little different than what everybody else is doing, that's OK.

Chapter 9
Ways to Miss

"Repetition is the mother of skill as long as there is skill in the repetition."

Neil Whyte

Watch any team warm up and there appears to be an endless number of ways to miss shots. But when you break it down, there are only four: left, right, long or short. You can have combinations of these, like missing long and to the left, but to make shots you need to send the ball straight and have the right arc.

Long or Short

Missing long or shooting an airball encompasses an incorrect application of force for your distance from the basket. You must launch the ball with the correct arc and the right amount of speed to have it land in the basket. Launch angle and launch speed are the two factors at work here.[1] Launch angle is shooting the ball up and forward to obtain the arc necessary for the ball to enter the rim. Launch speed is applying the correct amount of force to achieve the arc needed.

Obtaining the correct arc can be achieved by proper placement of your fingertips on the back of the ball at the moment of release. If your fingers are too high on the backside of the ball, you will have a flat shot and the ball will hit the back of the rim and bounce out. When your fingertips are too far under the ball at release, you will shoot an airball.

Nailing the distance aspect of shooting entails coordinating both arc and distance, so they are in sync. With experience, players quickly learn to get this aspect of shooting right because feedback is evident. Players shooting air-balls intuitively learn to make the necessary adjustments.

Players who back-iron shots are slower to make the connection, but this is an easy fix as their fingers only need to be slightly lower on the backside of the ball at release.

Left or Right

Watching how the ball contacts the rim and net will clue you in as to what happened during the release. You gain information from watching how the ball interacts with the basket—not watching the flight of the ball but watching how the ball hits the rim or net. This information provides you the feedback you need to improve.

Can you ride a bike or a skateboard? Most everyone learns to balance on these two items after a period of practice because feedback is swift and immediate. Fall off and you quickly realize you must do a better job of maintaining your balance. Over time, you improve your sense of balance and have no problem staying upright.

You can utilize the same concept when shooting a basketball. You are balancing the ball in your hand. If the ball becomes unbalanced during the shot, you will miss to the side. Left or right misses are the result of the ball becoming unbalanced during the wrist snap. Understanding this concept will provide you information as to what happened at the moment of release. You have feedback. Granted, it is not as attention-grabbing as falling off a bike, but a miss to the side indicates an unbalanced ball at the moment of release. You correct by doing a better job of balancing the ball in your hand through the entire shot.

Squirt

Missing to the left or right is caused by one of two things:

- Squirt—the fingers are moving toward the hoop, but the center of the ball is not

- The last moving joint and the control finger missed the centerline of the ball and basket at the moment of release

"Squirt" is a term used in billiards to describe the tendency of the cue ball to go off the aiming line. Whenever we hit the cue ball with a stroke that is off center, it is vulnerable to sliding off the tip of the cue stick. The

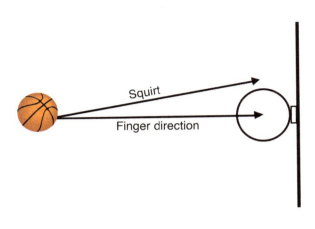

(Diagram by Justin Smith)

physics of pool and the physics of shooting a basketball are the same. Where you apply force, either with a cue stick or your hand, dictates the ball's direction.

Squirt happens when the ball skews to one side (this could also be called deflection). In billiards, this occurs when the cue tip is too far from the centerline of the ball and pocket as the ball separates from the cue stick. When shooting, the same concept holds true: squirt happens when the application of force is off-center of the centerline of the ball and basket at the moment of release. This action is a frequent occurrence because the size of the basketball increases the difficulty of controlling centerline.

We make shots by either applying force straight at the hoop or applying force through the centerline of the ball and basket at release.

Why We Miss

Vice versa, we miss shots because our fingers were not going straight toward the hoop, or our last moving joint and control finger are off-line of the centerline of the ball and basket. With either problem, we aren't controlling the centerline of the ball at the moment of release.

Everyone is aware our fingers not going straight can be a problem. The instructions "Hold your follow-through" and "Put your hand in the cookie jar" are designed to help increase your awareness of the need for your fingers to go toward the basket as you shoot. In actuality, holding your follow-through after the ball leaves your fingertips does nothing. Once the ball clears your fingertips, the shot is *over*. The train has left the station. It is what is happening at the moment of release which is critical. Instead of focusing on what your fingers are doing *after* the ball leaves your fingertips, notice how the ball contacts the rim and net. Evaluate what caused the ball to do what it did. Was it because your fingers weren't moving toward the basket at release *or* did you perceive the weight of the ball shift as you shot? (A slight shifting of the ball causes the ball to squirt and miss to the side.)

(Dziurek/Shutterstock.com)
Note how this player has three fingers primarily on one side of the ball. This finger position is susceptible to the ball squirting to her left.

Watching and focusing attention on how and where the ball hits the rim and net will provide clues to what you did at the moment of release. Those clues will give the feedback you need to improve your next shot. Focusing on your follow-through is a waste of time and a lost opportunity to learn from your mistakes.

Control Centerline

We know if you control the center of the ball, concerning the middle of the basket, at the moment of release, you will have a straight shot. This action is simple physics. The application of force through centerline results in the ball continually moving straight, regardless of what position your body is in or how you choose to apply the power. The ball doesn't care. It is going to travel in the direction which the application of force dictates.

So, the question is: How do you eliminate squirt?

If you pay attention, you can recognize this happening. You will perceive the weight of the ball shift slightly, right before release. To distinguish this better, square up to the basket, balance the ball in your hand, look at the rim, close your eyes and shoot. Feel the weight of the ball as your hand moves up and forward. With your eyes closed, your senses are enhanced and you can feel the weight of the ball shift better. Fine-tune your ability to keep the center of the ball balanced at release because this results in a straight shot every time.

Key Takeaway

When you control the center of the ball with the middle of the basket at release, the ball will go straight all the time. Nothing else matters. The important thing is to control centerline. Accomplish that and every shot you take will invariably go straight.

Chapter 10
Controlling Momentum

"Now that is what we call a high degree of difficulty—MISS!"

Hubie Brown, after watching a player throw up a shot

Years ago, if you were sitting next to me watching two high schools play a tightly contested tournament game, you would have seen an all-league point guard blow the game-winning layup. The coach had called timeout after the other team had turned the ball over. With a few seconds left, holding a one-point lead, the opposing team set up its full-court press. It didn't make sense to me, but that's what they did. The team behind ran a full-court play which worked brilliantly—because it resulted in their all-league player racing ahead of the field and shooting an uncontested layup.

The only problem? He missed.

Speed Creates Momentum

He was moving with such speed the ball bounced off the backboard, hit the rim and bounced off. I watched in amazement. How could an all-league point guard miss that shot? Didn't he know how to control momentum? Isn't this something skilled players pick up on their own? It wasn't like this player lacked skill. Against us, he scored 34 points.

The problem is kids rarely practice layups at *full speed*. Watch layup lines before the start of a game and you will see players approaching the basket at half-speed as they attempt their layups. The only time they shoot a full-speed layup is in game situations.

Controlling Momentum

How do you negate speed when attempting a breakaway layup? How do you cope with momentum when you are flying down the floor racing ahead of your opponents?

When you shoot a three-point shot or a free throw, you need your hand behind the ball applying force to get the ball to the rim. With layups, the speed at which you are approaching the basket provides momentum to the ball. Moving at half speed, you can have your hand directly under the ball and the ball will move forward and bounce softly off the backboard. The ball bounces off the backboard much harder when you are running full speed. The faster moving ball off the backboard regularly results in a miss.

How do the pros make full speed layups? *By moving their hand to the front side of the ball at the moment of release.*

(Aspen Photo/Shutterstock.com)   (Aspen Photo/Shutterstock.com)
Note the forward positions of the fingers on these layups.

Underhand Only

Young kids learn there are two types of layups: the overhand layup (hand behind the ball) and the underhand layup (hand under the ball).

Little kids use the overhand layup with the hand pushing the ball because they lack the power to get the ball up to the basket. However, as they progress and grow taller, strength is no longer an issue. The issue becomes momentum.

What changes from youth leagues to junior high to high school to college to the pros? The speed of the game. Everything happens faster. On layup attempts, players are approaching the basket with more speed. To manage momentum, they must use an underhand layup. Advanced players never have their hands behind the ball on layups because they are moving too fast. When they let go of the ball, it continues to move forward at the player's speed. Momentum is why skateboarders can flip the board while they are airborne and land on it; the skateboard continues to move forward at the same amount of speed as the airborne skateboarder. The basketball does the same thing; it continues to move forward at the same rate of speed as the moving player.

Easy Fix

The faster you are moving toward the rim, the more your fingers should move to the front of the ball at release. Controlling the ball from the front compensates for momentum. If you are traveling extremely fast, you will be looking at the palm of your hand when you release the ball. "See your palm!" was my reminder to players.

Practicing Layups at Game Speed

(Aspen Photo/Shutterstock.com)

Here is a simple exercise to practice full-speed layups.

Have three lines of players about 10 feet apart on the baseline. To start the drill, the player in the middle should move forward to the free throw line. The player in the middle has a ball and will be dribbling the length of the court and attempting a layup. The two outside lines are chasers and their job is to catch up to and reach with their inside hand to deflect the ball forward and away. They are trying to avoid the layup from happening without fouling. On the whistle, they take off and it is a race. Sometimes the ball gets tipped and the layup is averted. However, the layups attempted are game-like layups.

What Players Learn

The dribbling player soon learns to use only three or four dribbles to travel the length of the court. Players should push the ball out in front of them and run to keep up with it.

Another thing players learn is to not slow down at the last moment. The natural tendency is to slow down when you get close to the basket (especially at the younger levels). This moment is a prime time for the ball to be batted away by the defense.

To maintain your pace and not slow down, you must adjust to counter momentum. Move your hand position to the front of the ball. The speed of your forward movement will allow you to control the ball with your fingers in front. Controlling the ball from the front will enable you to apply whatever force is necessary to negate your forward momentum. When moving extremely fast, your hand and fingers will be on the front side of the ball at the moment of release. The photo of Ortiz shooting a layup demonstrates this well.

(Dziurek/Shutterstock.com)
This shot is a full-speed layup. Note how her fingers are in front of the ball to neutralize her momentum.

Also, from the defensive aspect, your players are learning how to back-tip the ball. With the experience gained from this drill, they should look for opportunities to do this in a game.

If you're the coach, you know your kids and it is permissible to give some a more significant lead if necessary. You want all your kids to have an opportunity to practice full-speed layups.

This drill works in a half-court setting as well. This exercise was probably the most game-like drill we did. The players enjoyed it and so did I.

Key Concept

To control momentum, the faster you are going, the more your hand should move to the front of the ball.

Chapter 11
Sidespin

"Don't confuse me with the facts; my mind's made up!"

Another item John Fontanella covered in his book *The Physics of Basketball* was how the spin on the ball affects a bounce pass.

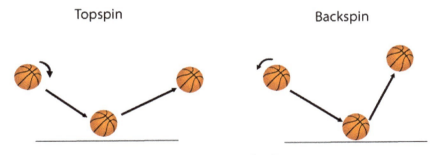

(Diagram by Justin Smith)

With topspin, the forward spin on the ball causes the ball to grip the surface of the floor and jump forward, gaining extra distance. With backspin, the ball bounces up and higher, with less horizontal distance covered. Also, Fontanella points out the spin on the ball after the bounce is faster with topspin than it is with backspin.[1]

A Shooter's Touch

All great shooters have what is called a "shooter's touch." This touch results in a favorable bounce (or bounces) when the ball contacts the rim. Fontanella makes the case a shooter's touch is the result of the slowest moving ball as it nears the basket, thereby providing the optimum chance of the ball going through the hoop.

A slow-moving ball as it nears the basket results in the ball colliding with the rim with less force. On contact, the ball bounces off with less energy, increasing the odds it will fall through the hoop.

Slowest Moving Ball Off the Backboard

In the same way as bouncing the ball off the floor using backspin to create a slower-moving pass for your teammate, you can use backspin off the backboard to establish a slower-moving ball as it nears the rim. Topspin off the backboard results in the ball moving horizontally across the face of the backboard. It covers more distance and has more spin than when you use backspin. Backspin slows the rate at which the ball turns and more importantly, slows the speed of the ball.

I didn't know how to do this when I was playing. I wish I had. Only after reading Fontanella's book and evaluating how spin affects the ball when it contacts the floor did I realize this is a great way to utilize physics to make bank shots.

Backspin on a Direct Shot

You create backspin on a direct shot by moving your fingers down the backside of the ball during the wrist snap. Snapping your wrist causes your fingers to move forward and down, which results in the ball spinning backward. Backspin is desirable because it increases the chances of the ball going through the hoop. When this backward-spinning ball contacts

(Daniela Parra Saiani/Shutterstock.com)   (Levante Media/Shutterstock.com)
Note the hand position starts behind the ball. As you shoot, you should rotate your hand to the front of the ball. Contact with the backboard slows the ball's speed.

the rim, friction causes the ball's forward movement to slow. A slower-moving ball at the basket gives you the optimum chance of making the shot.

How to Create Sidespin

Creating backspin off the backboard is different. Let's assume you are moving toward the basket from the wing area on the right side of the floor and you are going to attempt a driving ten-foot shot with your right hand. Instead of your fingers moving forward and down, as they would for a direct shot, your hand and fingers should relocate around and forward to the front side of the ball. To further clarify this action: when you start this ten-foot shot, your hand is behind the ball. As you shoot, bring your hand and fingers to the right side of the ball. Continue in a semi-circle around to the front of the ball before the moment of release. Done correctly, during the shooting motion your palm will rotate in a semi-circle around the ball.

For a simplification of this movement do this:

- Hold your hand up so you are looking at the backside of your hand, with palm facing away

- Rotate your hand so your palm is facing you

This action is the movement. Do this while you extend your arm. Use your forearm to rotate your palm from facing away at the start of the shot to facing you at the end.

You will know if you are doing this right by watching the ball come off the backboard. You will see the ball slow down and the spin reduced after making contact with the backboard.

(Nejron/Shutterstock.com)
Note how this player rotated his hand to the front of the ball.

Do What the Pros Do

When using this technique, notice how—when the ball contacts the rim—it seems to be pulled into the basket rather than bouncing off.

When you use this technique, you are using physics to help you make shots.

Lebron James is a master of this technique. So is Kawhi Leonard. It is something they have intuitively picked up through years of playing. They know it increases their chances of making the shot, which is why they do it.

You can do it, too.

Chapter 12
Bank

"It's a casino. The bank is always open."

Wayne Nelson

When he banked in the first one, I thought he was screwing around. "Why would he waste a practice shot during a competition?" I wondered. Then he did it again. Halfway through the semifinal round, watching ball after ball bang against the backboard and drop through the net, I realized what he was doing. With each shot, the backboard recoiled when the ball slammed into it.

"This guy is a genius!"

I was thinking out loud, but a newscaster immediately thrust a microphone in my direction.

"Do you really think so?" he asked.

"Absolutely!" I answered. "The portable backboard is flimsy, so it absorbs the ball's energy. All he has to do is throw it straight and the ball is going to drop through the rim."

It was the perfect strategy for the situation. Mounted backboards are stable, so the ball comes off with a lot of energy—that is, faster. Portable backboards recoil and absorb the ball's force, so the ball comes off slower. By banking his shots, Wayne Nelson of Smithville, N.J., was shooting at a larger target than his competitors. Nelson rode that strategy to second place in The Bogarda Free Throw Tournament in 2015 and walked away with $6,132.[1] Can you use physics to increase your shooting percentage? Wayne Nelson did. Although others could have shot bank shots, Nelson was the only one with the smarts and moxie to do it. By doing so, he was shooting at a more significant target than his competitors.

Advantage, Nelson.

Studying the Bank Shot

Rob Fanning, the producer of my video, *Secrets of Shooting*, thumbed through the material in the notebook I had handed him. A quizzical look came over his face. "We can't use this! This stuff belongs in a book."

In spring 2008, I did a study of the best contact area for bank shots from different spots on the floor. I used a transit to determine various angles to the basket and placed tape on the floor to designate different angles and distances. I marked the backboard to pinpoint the precise spot the ball made contact. I videoed attempts from each position. I analyzed hundreds of shots and noticed a pattern.

The angle from where you shoot determines the target area on the backboard. In other words, where you are on the floor determines where you should aim at on the backboard. It varies. The top corner of the square is where you want the ball to hit if you are at a 45-degree angle, but it is a guaranteed miss if you are shooting from below the block.

Contact Area on the Backboard

Dividing the court into sections provides a general guideline for where the ball should contact the backboard:

- Below the block, the ball should contact the backboard outside and above the marked square on the backboard
- Between the block and the first marker, the ball should contact the top upper corner of the square
- In the lane area, the contact area is inside the marked square on the backboard

These areas are where you want the ball to *hit* the backboard; it is not where you aim. Your target spot is inside the contact area. Since the ball is round and it is approaching the backboard at an angle, the outside of the ball will make contact with the backboard before the center of the ball does. Therefore, the aim spot and the contact spot are two separate places on the backboard.

This information served as a valuable teaching tool. It proved to be quite helpful. Players who struggled with bank shots before started making them.

Study Confirms Value of Bank Shots

In 2011, North Carolina State University conducted a study on bank shots which confirmed my findings and provided additional information. Engineers Larry Silverberg, Chau Tran, and Taylor Adams found at many angles inside 12 feet; bank shots can be 20% more effective than a direct shot.[2]

Aim Points and Contact Points of Bank Shots

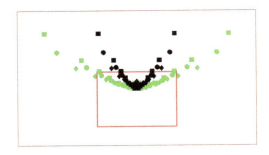

Aim points (black) and contact points (green)
$r = 13.75$ ft (square), $r = 9.842$ ft (circle), and $r = 5.905$ ft (diamond)

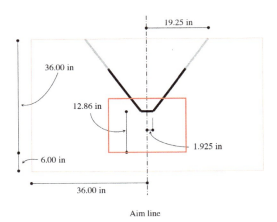

Aim line

*(Courtesy, Dr. Larry M. Silverberg,
North Carolina State University)*

    The top illustration shows the optimal aim points (black) and points where the simulated shots hit the backboard (green). The bottom illustration shows the aim line (dotted line) that can be crossed with the aim points to find the optimal aim points for a bank shot.

This study also concluded bank shots are more effective from the wing area and straight-on shots from the free-throw line area beyond 12 feet weren't as effective.

(Goran Bogicevic/Shutterstock.com)
Bank shots have a significant area of the rim to pass through.

Target Area versus Contact Area

These researchers identified not only the contact area of the ball but also the target area you should aim at.[2] A bank shot's target area is different than the contact area because the outside of the ball hits the backboard before the center of the ball does. Silverberg et al. showed the optimal aim points make a "V" shape near the top center of the square. Outside the free-throw lane, your aim points on the backboard are higher and farther from the rim. Inside the free throw lane, your aim points are lower and closer to the basket.

Takeaways

You gain a significant advantage by using the backboard on shots within 12 feet of the basket and where you are on the court determines your target area on the backboard.

One final thought: When using the backboard, never look at the rim and try to calculate the angle off the backboard. Look at the backboard, pinpoint your target area and aim for that spot.

Chapter 13
Free Throws

"While we are practicing we must know what we are doing: otherwise we will waste most of our time."

Gyorgy Sandor

Elena Delle Donne is one of the best free-throw shooters in the world. She currently has a career free-throw percentage of 94%, which is more than 3% higher than Steve Nash, who holds the current NBA record of 90.43%. In 2015, Delle Donne shot 95% for the season.

How does she achieve this high level of accuracy?

Delle Donne's Routine

She credits her seven-step routine:[1]

- Line up the top of her right foot on the nail hole
- Take three dribbles
- Place the index finger on the inflation hole
- Make a 90-degree angle with the right elbow
- Bend the knees

(Debby Wong/Shutterstock.com)

- Lift, flick and pop (lift elbow, flick wrist, pop ankles—but no jumping)
- Tell yourself it's going in

While this routine has worked well for Elena Delle Donne, it wouldn't necessarily work as well for someone trying to copy her method. Why? Notwithstanding the edge Delle Donne has experience-wise, this routine isn't the best for beginners because of its internal-based focus.

"Take three dribbles and place the index finger on the inflation hole"; this concentrates your attention on the ball, which is favorable because this is an external focus.

"Make a 90-degree angle with the right elbow; bend the knees; lift, flick and pop." This instruction has nothing to do with whether the ball goes in or not. Worse, they focus your attention on your body, which is an internal focus and has no place in a free-throw routine. We simply aren't adept at micromanaging our body movements. We do better focusing on what we are trying to accomplish because this frees your mind to do the work necessary to perform the task at hand.

"Telling yourself it is going in" is excellent because this creates a visual image of the ball going through the basket. Making the shot is what we are trying to accomplish. Visualizing the ball going through the hoop focuses your mind and allows your body to do what it has been trained to do.

What Is Important

Delle Donne's routine fails to mention the most critical element of shooting. Delle Donne excels at controlling the exact center of the ball at the moment of release. She eliminates the ball's tendency to squirt.

How does she do this?

By closing her fingers around the ball while she shoots. Closing her hand inward centers the middle of her hand in the center of the ball, which places her middle finger in the exact center of the ball at release.

Delle Donne brings her thumb forward while she shoots so her thumb and little finger provide excellent control and funnel the center of the ball to her index and ring fingers. These fingers then funnel the center of the ball to her middle finger, which leaves the ball last. By bringing her thumb forward, she is moving the exact center of the ball into the exact center of her hand. The ball then travels up the low of her hand to the middle finger.

With the index and ring fingers providing an even force on both sides of the ball, the ball is going to go straight all the time. The middle finger should continue an even application of pressure; applying additional force may cause the ball to squirt. I have mentioned this before, but this concept is extremely critical—when you control centerline of the ball, it goes straight because this is physics. Physics always works.

Excelling at controlling the center of the ball at release is why Elena Delle Donne makes such a high percentage of her free throws.

(Pavel Shchegolev/Shutterstock.com) Note how the fingers have moved inward during the shot. Closing the fingers together guides the ball into position, so your middle finger is in the center of the ball at release.

Two Types of Free Throws

It is common to hear the claim that all free throws are "mental." The truth of the matter is there are two types of free-throw situations:

- Routine free throws shot during the game
- Pressure free throws when the game is on the line

When you shoot free throws early in the game, any pressure you might experience is primarily self-imposed. People may be watching, or they may be more concerned about getting in line for popcorn. Free throws at this juncture are a token part of a game. You shoot free throws knowing it will help your team if you make them, but by no means is it going to be the determining factor in whether the game is won or lost. There is too much game left to play out. Making or missing at this point in the game is strictly a matter of technique or lack thereof.

However, at the end of a close game with the game on the line, it's a different story. Picture this: With one second left on the clock and your team down by one, you get fouled while shooting and awarded two free throws. The free throws you're about to attempt are going to either win or lose the game (or send it to overtime). Nobody is thinking about popcorn *now*. The game has their full attention and their focus is on you.

Now *that's* mental. Specifically, this is a situation where your mental state can interfere with the physical aspect of making a free throw.

(Aspen Photo/Shutterstock.com)
Note the time and score. This free throw might affect the outcome of this game.

Choking Study

In a 2014 experiment on choking, Vikram Chib, assistant professor of biomechanical engineering at the Johns Hopkins University School of Medicine, focused on the ventral striatum, a small area of our brain which processes reward but is also known to help control movement. According to this study, performance under pressure situations depends on two factors: the framing of the incentive in terms of loss or gain and a person's aversion to loss.[2]

The study showed people react differently. Those with high loss aversion choked when told they stood to gain a lot, while those with low loss aversion choked under the pressure of large prospective losses.

"We can measure someone's loss aversion and then frame the task in a way that might help them avoid choking under pressure," says Chib.

This study shows the way we frame the situation can affect how well we perform in a pressure situation.

Focus on the Process

What practical steps can we take to put us in a position to be successful?

In other words, how do you avoid letting the mental interfere with the physical? Or as some golfers would say, how do you prevent the yips?

Great question. The line starts here.

It is easy for me to remember a time when I was shooting a pressure free throw. Time seemed to stand still. When the referee eventually handed me the ball, it felt as if it weighed 10 pounds. I'd prefer to say I was as cool as a cucumber and nailed that free throw, but that wasn't the case. I blew it.

Why? Because I was concerned about how I was going to be perceived afterward and I let those thoughts interfere with my performance. I became more focused on the outcome than I was in the process. If Dr. Tom Amberry has told me once he told me a hundred times: "Focus on the process, not the outcome."

Easier said than done.

Coping with Pressure

When you encounter a pressure situation, there are a few things you can do to help keep your mind focused on the process.

First, what *not* to do: Don't immediately walk to the line, take your position and wait as I did. I stood in place at the free throw line and waited for everyone to line up and the referee to throw me the ball. I had plenty of time to think. In this situation, too much time to think can be detrimental. Plus, inaction doesn't help your muscles stay loose, which might explain why the ball felt like it weighed 10 pounds.

A better tactic is to keep moving. Stay in motion. Shoot one free throw and back off the line and step back to it when the referee is ready to throw you the ball. Moving helps keep you loose.

Deep Breathing

It also helps to take deep breaths. Science has proven you can control your body by how you breathe.

Esther Sternberg is a physician, author of several books on stress and healing and a researcher at the National Institute of Mental Health. She says rapid breathing is controlled by the sympathetic nervous system.[3] It's part of the fight or flight response, which isn't what you want to be experiencing while you attempt a potential game-winning free throw.

Sternberg says slow, deep breathing stimulates the opposing parasympathetic reaction—the one that calms us down. "The relaxation response is controlled by another set of nerves, the main nerve being the vagus nerve," she says. "Think of a car throttling down the highway at 120 miles an hour. That's the stress response and the vagus nerve is the brake. When you are stressed, you have your foot on the gas, pedal to the floor. When you take slow, deep breaths, that is what is engaging the brake."

In a paper published in the journal *Science*, researchers led by Mark Krasnow, a professor of biochemistry at Stanford University, found breathing can have a direct effect on the overall activity level of the brain. "This liaison to the rest of the brain means that if we can slow breathing down, as we can do by deep breathing or slow, controlled breaths, the idea would be that these neurons then don't signal the arousal center and don't hyperactivate the brain. So, you can calm your breathing and also calm your mind," Krasnow says.[4]

A Breathing Routine That Works

Breathing can change the mind, or the state of mind. Here is a simple step-by-step breathing method that works:

- Inhale deeply through your nose, expanding your stomach for a count of four—one, two, three, four
- Hold that breath for a count of four—one, two, three, four
- Exhale completely through your mouth—one, two, three, four
- Hold the empty breath for a count of four—one, two, three, four

Be sure to inhale through your nose while expanding your stomach. On the exhale, contract and empty the abdomen of breath. Deep breathing works for more than shooting free throws in a pressure situation. Breathing in such a manner can be used to calm your body before any stressful situation; like speaking in public.

Have a Free Throw Routine

A familiar ritual will help in these situations. Routines help you shoot a higher percentage, according to a study of NBA players, and assists in pressure situations as well. Focusing on your routine keeps your mind occupied. (However, this is easier for males than females. One of the many differences between the male and female brains is men focus on one thing at a time while women have a better ability to multi-task than men.[5])

The best free-throw routine I have ever seen is Jerry West's routine, although I also admire the run-through shooting motion that Steve Nash did.

Best Routine

What did Jerry West do?

He took his position at the line, bounced the ball three times and shot.

What was special about that? It was where he bounced the ball. West dribbled the ball in the same vertical plane as his shot line. Looking at it from a different perspective, he

(Aitor Bouzo Ateca/Shutterstock.com)
Note how the player is running through his shooting motion without the ball. Mimicking his shooting motion was part of Steve Nash's routine and it must have helped. Steve Nash retired with the highest career free-throw percentage in NBA history with 90.43%.

shot the ball into the floor three times before he shot it at the basket. He dribbled with the same motion as when he shot. His dribble motion and his shooting motion were identical. By doing this, you can practice your wrist movement and finger placement on the ball. When you shoot, use the same wrist motion and finger placement as when you dribbled.

Try it. It works.

"I Am Excited"

Framing how we view ourselves in a stressful situation matters. Rather than focusing on your nervousness, tell yourself, "I am excited."[6] How we talk to ourselves is important and this creates a better mindset than saying, "I am nervous" or "I am calm." By telling yourself you are excited, you are acknowledging the truth of the matter while creating the right frame of mind for an excellent performance and a successful outcome.

Familiarity Helps

If you miss a crucial free throw, don't despair. Know that having been through that experience will enhance your chance of being successful the next time. The more we experience a new or different situation, the more we become accustomed to dealing with it.[7] We adjust and the new and different becomes the norm. The more it becomes the norm, the lower your anxiety will be. Learn to look for and take on new challenges because they will help you grow.

Chapter 14
Lateral Movement

"To go where no one has gone before."
Samuel Peeples

Lateral movement is any movement which is moving sideways to the basket. Shots taken while moving laterally are a regular part of the game. A superb example of a lateral moving shot would be Kareem Abdul Jabbar's legendary sky-hook. He shot this while moving at an angle to the basket. (Videos of him making this shot are available on YouTube.) Making shots while driving at an angle to the basket requires that you compensate for momentum. The same concept used to shoot layups is in effect when shooting while moving laterally. To make any moving shot, players have to correctly account for how their speed is affecting the ball's momentum.

Stationary Shots

When you are standing still, the alignment at release is:

- Low of your palm
- Fingertips
- Center of the ball
- Center of the basket

You are guaranteed a straight shot when you have these four items in the same vertical plane at the moment of release.

(Natursports/Shutterstock.com)   *(Natursports/Shutterstock.com)*

Klay Thompson provides an exceptional example of how to make shots while moving laterally. With his palm centered, he has complete control over the ball. At release, his fingertips are on the front side of the ball to cancel out his momentum.

Moving at the Basket

As we discussed in a previous chapter, the way to negate forward momentum on speed layups is to move your fingers to the front side of the ball at release. The faster you are moving towards the basket, the more your fingers should move to the front of the ball to offset your momentum. This finger position at release will produce a slower moving ball at the rim.

Moving Laterally to the Basket

The same concept applies when shooting while moving laterally to the basket. Controlling your momentum requires your fingertips to leave the ball from the front side. How far your fingers will be in front of the center of the ball will be determined by how fast you are moving.

With practice, you can learn to sense the adjustment you should make by what area of your fingertips are in contact with the ball at the moment of release. When stationary, the contact area of your fingers at the moment of release will be slightly closer to the ends of your fingertips than when you are moving laterally.

Chapter 15
Downtown

"The finger has hundreds of sensors per square centimeter. There is nothing in science or technology that has even come close to the sensitivity of the human finger, with respect to the range of stimuli it can pick up. It's a brilliant instrument. But we simply don't trust our tactile sense as much as our visual sense."

Mark Goldstein, a sensory psychophysicist

For outside shooting, utilizing a physics-based perspective, there are three criteria you must meet. They are:

- Control centerline of the ball and basket
- Apply the appropriate amount of force to the ball for where you are on the floor
- Correlate the correct arc with your distance from the basket

What makes long-distance shooting difficult is the margin for error is less than it is for close shots. The bandwidth to sink a two-foot shot is more extensive than the bandwidth to drain a 20-foot shot. In other words, the farther you are from the basket, the less tolerance you have. Long-distance shooting requires superior control of the exact center of the ball at release.

There are two distinct techniques for maintaining control of the center of the ball at release. Both methods work well and are used extensively by players. The two ways are funneling and wedging.

Funneling

As we covered in an earlier chapter; Curry, Thompson, Harden, Lowry, Leonard, Redick and Irving funnel the ball to the middle finger at the moment of release. They move their thumb forward while they shoot, so the ball becomes centered between the index and ring fingers. Then they use the index and ring fingers as a funnel to direct the exact center of the ball to the middle finger. This funneling action serves to apply force in the exact center of the ball at the moment of release.

Wedging

Wedging is a term to describe the action of centering the middle of the ball between two fingers. Historically, the index and middle fingers were taught to gap the center of the ball. However, the ring and middle fingers can also be used to control the middle of the ball. This release provides excellent control and is very biomechanically-friendly. The key to wedging is to position your fingers to place the center of the ball between them and to apply equal force with both fingertips.

Finger Length

Which release you choose to use may be determined by the length of your fingers. Scientists have noticed that men's ring fingers are generally longer than their index fingers. With women, their index fingers are usually longer.[2] They've called this difference in length between the index finger and ring finger the "2D:4D ratio." This term references the second digit (index finger) to the 4th (ring finger). Varying finger lengths will influence which release provides the most control on an individual basis. Find what works best for you and use it.

Palm Shooters

Thompson, Durant, Harden, Lowry, Leonard, Curry, Redick and Irving are palm shooters. When they start their shots, the ball is sitting firmly down on their palm. This starting position allows the hand and fingers to control more area of the ball. As the shot progresses, they move their fingers forward, which keeps their thumb and fingers on the ball a split-second longer.

 This approach is different than old-school shooting instruction. Back in the day, shooting instructors taught players to get the ball up off their deep palm and onto their fingertips. They reasoned the ball sitting on your palm would reduce backspin, but that is not the case. The ball may start the shot on your palm, but it quickly moves to the fingertips, so the final result is the same.

Players have learned that it is helpful to use the palm to generate the initial force to the ball when shooting. Fingers will distort and change position but your palm will not.

The "L"

Years ago, shooting instructors insisted players make an "L" with their upper arm and forearm. (One exception was Andy Enfield, now the coach at the University of Southern California.) The idea behind this teaching is players should lift the ball before pushing it.

Making an "L" does restrict movement. But this restriction results in a loss of power. This idea originated in the 70s and 80s before the three-point shot existed. Limiting movement might be desirable when teaching 15-foot jump shooters, but today's players are launching from behind the arc. More power is required.

Another drawback to the "L" is it positions the elbow farther away from the body. The shoulder is a ball and socket joint and provides the most control to your hand when it is down and closer to your side. Remember how the surgeon loses precision when his hand is above shoulder level? You maintain more control when your elbow is lower than when it is above your head.

(Aspen Photo/Shutterstock.com)   (Marcos Mesa Sam Wordley/Shutterstock.com)
Note the "L" position of the upper arm and forearm.

Use the "V"

Making a "V" with your upper arm and forearm allows you to generate more power. Furthermore, the lower starting position of the ball provides more control.

To be an outstanding three-point shooter, you need to generate force efficiently. Shot-putters don't throw with their arm in the "L" position. You shouldn't either. To generate force as efficiently as possible, take a lesson from Stephen Curry. How does he hit half-court shots at such a high percentage? He gets his arm in close to his body and generates force efficiently. He makes a "V" with his upper arm and forearm—as do Kyrie Irving, Isaiah Thomas, Frank Mason and others.

(Aspen Photo/ Shutterstock.com) (Marcos Mesa Sam Wordley/Shutterstock.com)
Note the "V" of the upper arm and forearm.

Stay Loose

Tight muscles are slow muscles. Staying loose and relaxed will produce better results. Tim Burke has the record for the longest drive in a golf competition with a whack of 469 yards. His secret? "Being more relaxed and fluid will help produce more speed and the most efficient golf swing," he explains.[3]

Whether you are hitting golf balls or shooting from downtown, the same concept applies. Relaxed is better.

Chapter 16
Dr. Tom

Dr. Tom had a very successful podiatry practice. To promote his business, he had 50 suits tailor-made and embroidered with images of feet. The suits were eccentric and attention-grabbing. Dr. Tom deducted their cost on his taxes as a business expense. He got audited by the IRS and they questioned the legitimacy of the deduction. A meeting was scheduled and Dr. Tom wore one of the suits to the meeting. The head auditor looked him over and said, "I am going to allow it. Anybody who has the balls to wear something like this can have the deduction."

"Write this down!" the voice over the phone instructed. The urgency in his tone immediately made me grab a pen and paper. "Focus on the process—not the outcome!"

Unmatched Accuracy

Dr. Tom Amberry was the most accurate free-throw shooter ever. No one else is even close. In 1993, at the age of 71, he shot for 12 hours in a Seal Beach, California, gym and made 2,750 in a row. He would have made more, but as he said on David Letterman's show, "They kicked me out of the gym" because they had league games scheduled.

Dr. Tom shot from 7am to 7pm. Twelve hours. Three hours at the gym and I am ready to head home. And to not miss a single shot during a 12-hour span?

Impressive as that was, there is more to the story. Dr. Tom was a foot doctor. (As he said, "I told my Mom I wanted to be a podiatrist. She said, 'What is wrong with being a rich-diatrist?'") He retired at 69. He soon became bored, so he started shooting free throws. A natural left-hander, Dr. Tom shot so much he blew out a tendon in his left arm. So he switched to shooting right-handed. Dr. Tom made 2,750 free throws in a row shooting with his off-hand.

The rest of the story is Dr. Tom shot 500 free throws every morning. He would shoot 25 in a row and then stop to write down how many he made. Over 450 times, Dr. Tom made 500 out of 500 free throws.

World Traveler

A 71-year-old man making 2,750 free throws in a row received a great deal of attention. Dr. Tom was an inspiration. He traveled the world shooting free throws. He appeared on *The David Letterman Show*, *Ripley's Believe It or Not*, *That's Incredible*, *Monday Night Live*, CNN, ABC, *NBC Nightly News*, *Late Night* and dozens of other programs. He was invited by college coaches to give clinics to their players. Also, he became the free-throw coach for the Chicago Bulls for three years. In competitions, he never lost. Dr. Tom won 288 gold medals in Senior Olympic Games and won the 2002 World Masters Games' free-throw champion title in Melbourne, Australia. Dr. Tom consistently outshot NBA players. ("I beat Kobe twice," he proudly stated.)

Dr. Tom teamed up with author Philip Reed to write *Free Throw: Seven Steps to Success at the Free-Throw Line*,[1] and created an instructional video, *Make Every Free Throw*. They also established the website Freethrow.com.

Becoming Friends

I first became acquainted with Dr. Tom when I called him with a question about his instructional video. He was candid and I enjoyed talking with him. I sent him my video, *Secrets of Shooting,* and followed up with a call asking what he thought of it. "Using that kid was smart!" was his comment.

The phone calls continued and after I set my first record Connie and I visited him. Dr. Tom pulled out videos of appearances on various shows and regaled us with stories of his traveling the world shooting free throws.

The Amberry Hilton

Dr. Tom put us up in the spare bedroom of his Seal Beach, California, home on that and subsequent visits. We would get up and go to breakfast at 6 a.m. with him and Marcee Heltzen, his lively friend and companion.

The Primrose and Huff's were his favorite places to eat and oatmeal was his staple fare. Dr. Tom and Marcee introduced us to California's In & Out Burger and the great shakes at Ruby's Diner (he and I share a love of ice cream and we indulged it). The four of us had a great time together. With them, Connie and I laughed more than we ever have. They would regale us with stories of experiences they had together. For example, one story detailed the time they attended a high school basketball game. The place was packed. Time was running out in a close game when a player stole the ball and had an uncontested path to the basket. According to Marcee, it should have been an easy layup, but instead, he tried to showboat by "contorting his body all over the place." He missed what should have been the game-winning layup. "The gym went silent," Dr. Tom said. "You could have heard a pin drop. At that moment Marcee yells out, 'You *shithead*!'"

"Everyone turned around and looked," Dr. Tom chuckled. "God, I was so embarrassed. I couldn't get out of there fast enough."

"Yeah," Marcee chimed in. "I looked around and you were *gone*. And after that, he never would sit with me again, either."

Dr. Tom's One-Liners

Dr. Tom was the king of the one-liners. When he got rolling, he could rattle them off...

> "If love is like a burning flame—marriage is like a whole damn fire department."
>
> "When you get married, always get married first thing in the morning. That way, if it doesn't work out, you haven't ruined your whole day."
>
> "I am still missing my husband. But my aim is getting better."
>
> "Growing up, we were so poor I had to wear hand-me-down clothes. And all I had were two older sisters."
>
> "Marriage is like taking a bath—after a while, not so hot."
>
> "Statistics are like a bikini. What they reveal is exciting. What they conceal is vital."
>
> "It may be true you can draw more flies with honey, but who wants more flies?"
>
> "Man is not completed until he gets married. Then he is done."

"A poor politician is a poor politician."
"Her Daddy has a PhD.—Papa has dough."
"I gave my girl a ring and she gave me the finger."
"He was so rich; his wife had diamonds as big as horse turds."
"It is much better to want the mate you do not have than to have the mate you do not want."
"A man with a broken condom is often called 'Daddy.'"
"Why is sex like credit? Some get it and some don't."
"I never knew my daddy drank until I saw him sober."
"Better to lose a lover than to love a loser."
"Jokes are like sex. Neither's any good if you don't get any."
"Old age comes at a bad time."
"You may have snow in your hair and summer in your heart, but if you don't get some spring in your ass, we will be here until fall."
"I went to school to study art and chase girls. It wasn't long before I dropped the art."

When Dr. Tom turned 94, I asked him what it was like to be 94. "Well," he quipped, "I stopped buying green bananas."

Look for the Good

Dr. Tom's mantra was, "Look for the good." This saying was a credo he lived. Never would you hear him speak ill of anyone.

Dr. Tom was a *straight shooter*, literally and figuratively. (He called me that a couple of times, which was his ultimate compliment.) He loved to share what he knew and he did. We talked about shooting at length and he would get on me for being too technical. "Players aren't interested in all that stuff," he insisted. "You have got to keep it simple. KISS—keep it simple, stupid."

He listened with interest and took great delight in my successes. When I failed, he would remind me, "That is part of the process. You will learn from it and next time you will do better."

Dr. Tom was my coach, my mentor and my champion. He believed in me before I did. He was unceasingly positive and a constant supporter. Dr. Tom was my role model. I relied on him for advice whenever I encountered new situations, like going on *The Tonight Show with Jay Leno*.

Dr. Tom was my best friend. He passed away in March 2017 and I miss him greatly.

Dr. Tom's Seven Step Process

During our visits, I questioned him about how he could be so accurate. Dr. Tom had a seven-step process he used for each shot. It took him six seconds to run through the sequence:

- Feet square to the line
- Bounce the ball three times with the inflation hole up
- Put your thumb in the channel (groove) with your middle finger pointing at the inflation hole
- Elbow in
- Bend your knees
- Eyes on the target
- Shoot and follow through

More than Focus and Concentration?

Others who have tried to mimic his technique have improved but failed to achieve the spot-on accuracy of Dr. Tom. What did he do that other shooters didn't?

He would tell you it was his focus and concentration. There is no doubt he was the absolute master of that. But going a step further, could his accuracy be attributed to *what* he focused on?

External Focus

In Chapter 1, you learned of Gabrielle Wulf's studies of external versus internal focus and how external focus is better. Dr. Tom utilized external focus when he locked in on the inflation hole of the ball. Concentrating on the inflation hole played a crucial role in his seven-step process.[1] Dr. Tom believed in focusing on the inflation hole because it was the one thing all balls had in common and he used it to get his grip in the same position each time.

Hand Centered

From a physics-based perspective, it is worth noting his grip placed the center of his palm in the center of the ball. This hand placement provided him considerable control over the ball. With his thumb in the groove and his middle finger pointed at the inflation hole, his hand covered a large area of the ball.

More importantly, the middle of his hand aligned with the exact center of the ball. With his palm directly in the center of the ball, his lower thumb pad was on the left side of the ball and his hand pad (the outside edge of the hand) was on the right. His middle finger was in the center of the ball.

Here is the outline of Dr. Tom's hand on the ball. Note how Dr. Tom's palm and middle finger are in the center of the ball.

Square Stance

Dr. Tom advocated a square stance. This position is when both feet are equal distance to the rim.[1] Being square influences the movement of your hand toward the basket. By merely extending your arm, your hand will move straight toward the hoop. Try it. Stand square to the rim, with your hand by your side. Bring your hand up level with your chest and extend your arm. Your hand will go straight.

Efficient Shot

Dr. Tom generated power by using a lower starting point and he was a one-motion shooter. Once he started his shot, it was one fluid motion. Because of this, he didn't have to generate any additional force from his wrist snap; all he needed to do was dictate direction. Generating force efficiently helped improve his accuracy and provided more control over the center of the ball.

Consistent Hand Position

Dr. Tom used the same precise grip on the ball each time. His grip placed the middle of his hand in the exact center of the ball, while he extended his arm and snapped his wrist. Full extension of his arm squared his palm to the rim before he released the ball.

Wrapping Up

What can we learn from Dr. Tom?

Dr. Tom was a master of a straight shot because he virtually eliminated the ball's tendency to squirt. He controlled the ball with his entire hand. At release, his middle finger was in the exact center of the ball which resulted in a straight shot every time. In a nutshell, Dr. Tom controlled the center of the ball relative to the middle of the basket. The degree to which Dr. Tom was able to do this on such a consistent basis was incredible.

Chapter 17
Tall vs. Short

"It's not that I'm so smart, it's just that I stay with problems longer."

Albert Einstein

Dustin Ridder is one of the smartest guys I know. During his short tenure of working for the government, this became apparent to me one morning when he looked up from the training manual he was studying and stated, "They made a mistake in this equation."

I looked at him like he was nuts. "What do you mean?"

"Right here. They left [whatever it was] out of this equation. The answer should be [this] instead of what they have."

Now, I don't know about you, but when I read a training manual, I don't check equations. I assume they are correct. After all, they are teaching the material, right?

The second time this happened, in pretty much the same way, I looked at Dustin. "What are you doing here? Why don't you go get a real job?"

Which he eventually did. But before he left, one morning he looked over at me and said, "You know Bob, you are like the Einstein of basketball shooting."

Now, this was high praise coming from someone I respected and my chest started swelling along with my head size.

"Well, thank you, Dustin."

"Oh, I didn't mean you are smart," Dustin quickly clarified. "What I mean is you can stay focused on something for long periods of time. Like Einstein. He worked for 18 years before they verified his work. I can't do that. I study something for three months, but then I get bored with it and move on to something else."

"Oh." This clarification took the wind out of my sails.

"So, what you are telling me is I am kind of like the mentally-challenged who can fixate on mundane tasks for long periods of time?"

"Exactly!" Dustin beamed, obviously pleased I finally understood his comment.

Feeding the Misconception

While reading *The Art of a Beautiful Game* by Chris Ballard, it was my turn to notice a mistake. This book is excellent: well written and insightful. I enjoyed reading it and learned a lot. However, when it came to the chapter on free throws, Ballard stated it's more difficult for big men to shoot free throws. Ballard's interpretation of Fontanella's book, *The Physics of Basketball*, led him to the conclusion taller players have a smaller window to get the ball into the rim.[1] That hadn't been my take on Fontanella's point. So, I pulled out his book and looked at it again.

Arc for a Shooter's Touch

A slow-moving ball when it nears the rim gives you a "shooter's touch." According to Fontanella's calculations; to produce the slowest-moving ball as it nears the rim, a 6-foot player should launch the basketball at 51 degrees. The slowest-moving ball for a 7-foot player requires a launch angle of 48.7 degrees. Releasing from a higher point, the taller player has the advantage of using slightly less arc.

Visualize a window directly above the front of the rim through which the ball must pass to get into the basket. This window will vary in size for players of different height. When both players utilize their ideal arcs for the slowest moving ball as it nears the rim, the 6-foot player's window size is 13.4 inches. For a 7-foot player, the size of the window is 14.4 inches.[2]

Larger Target

If both players use their ideal launch angles, the one-inch increase in window size provides taller players a larger target area. Furthermore, since the taller player is releasing from a higher vantage point, the travel distance of his ball will be slightly less.

Validation

Questioning my take on the situation, I called Fontanella.

"As usual, you are right again," he responded. "This is one of the reasons I am considering writing a sequel, to help eliminate misunderstandings such as this."

Big Hands Misconception

Some commentators have attributed the poor free throw shooting of taller players to the fact that they have big hands. In basketball, huge hands are not a disadvantage—they are an advantage. Bigger hands enable better control over the ball. Connie Hawkins, one of my favorite NBA players as a teenager, had enormous hands and he could shoot. Bob Cousy, my childhood idol, credited part of his success to his large hands, which made it easier for him to control the basketball.

Controlling the center of the ball at release is easier to do when you have large hands. The trick is to use them effectively.

Tall Is Better

When you consider it from a practical standpoint, the taller player has a distinct advantage over the shorter player. To demonstrate this, kneel down and shoot. What you will notice is it takes more effort and ball speed at release to get the ball to the rim. Taller players can shoot slower and use less speed because by releasing from a higher point, the ball travels a shorter distance. This fact provides yet another slight advantage taller players have because studies have shown shooting more slowly helps increase accuracy.

(XiXinXing/Shutterstock.com)
Shooting free throws—who has the advantage?

Why Guards Shoot Better

Fontanella pulls the plug on the argument big men are at a disadvantage when it comes to shooting free throws. They have an advantage over shorter players. Which raises the question, why do some tall players struggle at the line?

My hypothesis is because guards handle the ball more, they develop better touch in their fingertips and therefore can control the center of the ball better. Guards practice dribbling more because they have to. Watch Stephen Curry's pregame warm-up. The number of dribbles he executes is astounding.

To dribble expertly, you must develop a feel for the ball. Extra practice dribbling results in guards developing an enhanced sense of touch. Greater touch boosts their ability to control the ball when shooting.

What is the key takeaway here? Improve your dribbling and you will improve your shooting. All good shooters develop an enhanced feel for the ball. Most know if they will make or miss the shot the moment it leaves their fingertips. Improving your sense of touch is developed and practicing dribbling will help you do it.

Chapter 18
High or low

"It is generally unhealthy to be overly concerned about what other people think."

David Gray

For the most efficient shot possible, it is helpful to be a one-motion shooter. Stephen Curry has popularized this technique. Curry is a one-motion shooter, which is different from what my generation learned.

Old School

Back in the day, shooting instruction advocated a two-motion shot:

- One: Raise the ball to forehead level
- Two: Shoot

(Aspen Photo/Shutterstock.com)

In the 1980s, Peter Brancazio, a physics professor at Brooklyn College, did a study in which he determined adding two feet to the height at which a shot leaves the player's fingers increases the success rate by 17%.[1]

This might have led to an increase in jump shots. What better way to release than from a higher vantage point?

With the jump shot, you were taught to shoot when you reached the top of your jump. This creates a bit of a one-two effect. One: Jump and bring the ball up to forehead level. Two: At the top of your jump, shoot. The advantage to doing this is you are indeed releasing from a higher point. The drawback is it's slower.

(Richard Paul Kane/Shutterstock.com)
A player shooting a jump shot. Notice he is reaching the apex of his jump and his wrist snap hasn't started.

Quick is Better

What has Stephen Curry proved beyond a shadow of a doubt?

Quickness rules. In today's game, how quickly you shoot outranks how high you release. Today's players are more athletic and taller than they were 50 years ago, not to mention faster and stronger.

Curry has the quickest release in the NBA. If you haven't watched the Sports Science segment on YouTube with Stephen Curry, I encourage you to do so. It reveals by the time most NBA players release the ball, Curry's shot is already 12 feet in the air because it only takes Curry 0.4 of a second to get his shot off. The average NBA player uses up 0.54 of a second to shoot. Curry gets his shot off quicker because there is no hesitation in his shot. The ball keeps moving. The ball doesn't slow down (or stop altogether) like it does when some players shoot.

How does he do this? Curry shoots on the way up. Jerry West did the same thing. They pull the trigger before they reach the top of their jump. The result is a quick shot which is virtually impossible to block.

Tough to Guard

Having a quick release makes blocking their shots much tougher. Another advantage to shooting on the way up is it uses the early momentum created with your lower body. If you watch players shoot and their feet fly forward before they land, you know they used more of the old one-two shooting style. Their feet moving ahead indicates they used more of their upper body to provide the bulk of the force required to power the shot. When you bring the ball up, hesitate and then shoot, your feet are going to move forward because you are applying force from a higher vantage point and the recoil will push your shoulders back, away from the basket. This technique is not the most efficient way to generate force.

Use the early momentum of the ball moving upwards and shoot on the way up. This act will allow you to release with a higher launch angle, which will increase the size of the rim area, so you are shooting at a more significant target. This technique is what Stephen Curry does so well. He quickly launches at a higher angle than average, which may explain why he is so accurate.

Mid-range

Not to say there isn't a time and place to use the one-two motion when shooting a jump shot. There is. It is the mid-range game. When you are in closer to the basket, your upward momentum makes it tough to shoot the ball on the way up without throwing it through the rafters. In this situation, releasing after you hit the apex of your jump provides better control than shooting on the way up. Since you are in close, generating the force necessary to get the ball to the basket isn't an issue.

But when you get behind the arc shooting a three-point shot, release the ball about the time your toes come off the floor. This action enables the efficient generation of force, thereby improving your accuracy.

Negative Motion a Negative

Negative motion when shooting is when you bring the ball away from the basket. An extreme example of this is when players start their shot from behind their head. Taking the ball to the back of your head to shoot is slower and results in less arc. Neither of these is helpful in game situations.

(Aspen Photo/Shutterstock.com)
Note this player has a "V" between his upper arm and forearm, which will help increase his range.

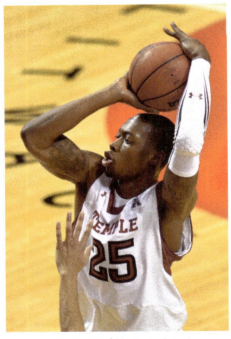

(Aspen Photo/Shutterstock.com)
Note how far back the ball is behind the player's forehead. This negative motion is not helpful for outside shooting.

Shooting from Distance

As a general rule, the farther you are from the basket, the lower you must start your shot.

When a player shoots with too high a release from behind the three-point line, I will have him shoot from half-court. They quickly learn to get the ball to the basket, they have to lower their starting point to the chest area or lower. Shooting from half-court resembles an end-of-the-quarter shot which is common for players to attempt during a game. This exercise has value in teaching this simple concept: The farther you are from the basket, the lower your starting point should be.

Distance Shooting in Action

The best example of this happened during my tenure at Onaga (Kansas) High School. We were playing in sub-state, and the opposing team scored with 2 seconds left in the first quarter. The ball was quickly inbounded to point guard Craig Boswell. He took one dribble and launched it from their free-throw line—a full three-quarter-court shot. *Swish.* The fans went nuts.

Now, that is not uncommon. You will occasionally see a long-distance shot during the season.

At the end of the third quarter, Boswell did it again. It was identical to the first quarter; one dribble, release at their free throw line. Nothing but net. That brought the house down. That is the only time I have ever seen two three-quarter-court shots made by the same player in the same game.

(Keeton Gale/Shutterstock.com)
Note the lower starting point of this player.

His technique was flawless. Boswell took one dribble and started the shot at his hip, using his lower body to generate the initial force and continue it right on up through his upper body.

What was impressive was how smooth Boswell made it look.

Chapter 19
Exercises

Do the work today, and you will be better tomorrow.

While in college, Pistol Pete Maravich was answering questions after a game. Maravich led the nation in scoring three years in a row, averaging 44 points per game. "How long did you have to practice to get this good?" the reporter asked. Maravich answered, "Eight hours a day in the summer and four hours a day in the winter. You don't get this good by wishing."

K. Anders Ericsson's work on deliberate practice recognizes the value of a teacher or coach who can provide exercises that will improve a player's performance.[1] An astute coach can also provide feedback. Feedback is a critical element for improvement.

If you don't have a great coach, focus on the three F's: focus, feedback and fix it.[2] Break things down into components you can repeat and analyze.

Over the last ten years, I have used many different exercises or drills. Here are a few of the best ones.

High Bounce, One-Hand, Shoot

Take the ball and slam it into the floor so it will bounce high enough that you can catch it with your shooting hand. Keep your off-hand off the ball as you control it. Shoot, only using your dominant hand.

You will have success with this pretty quickly because your shooting hand is centered and your off-hand doesn't screw things up. The key is to have your palm in the center of the ball relative to the middle of the basket at release. The movement or angle you use to get to that point makes absolutely no difference.

Continue to work on this, and with practice, you will not have to bounce the ball as high. Continue to shoot from a lower bounce. Give it some time; don't rush it. Work on it a little each day. You will progress to the point where you will be able to shoot off the dribble.

Eyes-closed Free Throw

Stand at the free-throw line and look at the rim. Shut your eyes and shoot.

Lock in on the weight of the ball while you snap your wrist, and feel your fingertips leave the ball. (This is easier to do with your eyes shut.) Open your eyes after the ball is on its way and watch where it goes. Notice where it hits on the rim. Any misses to the left mean you were applying more force on the right side of the ball at release. Vice versa, if the shot misses right, you crossed the center of the ball and were applying force on the left side of the middle of the ball at release.

It will take time to develop your sense of touch. Know that each practice session brings more sensitivity to your fingertips. Your brain will change and adapt to accomplish what you demand of it. Work on it for a bit and then try it again tomorrow. Stay with it. You can get this.

Two-Ball Dribble Against a Wall

Take two balls and dribble against a wall at head level or higher. Stand two feet away from a wall and dribble against the wall.

Try to dribble for 30 seconds without losing control. Build up to five minutes.

Shoot While Someone is Throwing You a Bounce Pass

Have someone throw you a bounce pass with a second ball while you are shooting.

This exercise makes you look down to catch the second ball. Ball-watching at the lower levels can be a problem because during games you speed up and look up more quickly than usual. This act results in your shot falling short. This issue is more prevalent at lower levels than with advanced players. Yes, Dirk Nowitzki, Stephen Curry, and other pros look up to watch the ball in flight, but they do it after the ball leaves their fingertips. Also, they look up with their eyes—they don't tilt their heads back as much as younger players do. Catching a pass immediately after you shoot curbs looking up to admire the ball in flight because you don't have time to do both.

Shoot Off the Side-edge of the Backboard

Shoot against the edge of the backboard and have the ball come directly back to you.

Distance Exercise (for advanced players)

Shoot from the free throw line. Have the ball make contact with the front of the basket before going in. Swish the next shot. Then bank it in. Repeat.

Deviation Exercise (for advanced players)

Shoot and have the ball graze the inside of the right side of the rim as it goes in. Then swish a shot. Then shoot and have it rub the inside of the left side of the basket as the ball goes through the hoop. Repeat.

Shoot while Moving Laterally (for advanced players)

Stand at one elbow of the free throw line and move laterally along the free throw line while you shoot. Focus on using the centerline concept. Using the centerline concept allows movement—you don't have to be stationary.

Solitary Knockout

To increase my speed, I played the game "Knockout" with myself. I used two balls. I shot the first one with a high arc and then grabbed the second one and shot it with a flatter arc, so it beat the first ball through the basket.

Try it sometime. It is a challenge.

Chapter 20

Closing Comments

*It is not a matter of doing your best. It is more a matter of
finding out how good your best can be.*

Jeff Liles lives in Kingfisher, Oklahoma and he invited me down to do a couple of school presentations. After observing them, Jeff offered this critique. "You want the kids to know everything you know," he said. "That isn't necessary. They don't need to know everything you know. Talk less and shoot more. Entertain them. Years from now they won't remember what you said, but they will remember the shots you made.

(Author giving a presentation for the Wichita East High School students.)

Besides, you only have their attention for ten minutes max. Talk less and shoot more."

Jeff was right. I took his advice.

Long Time Coming

This book has been ten years in the making. I started writing six years ago and realized there was much I didn't know, so I put it aside and kept trying to figure things out. Eventually, I fully grasped the centerline concept and some other insights pulled it all together. Now it makes perfect sense.

In writing this book, I kept Jeff's advice in mind. It is not essential that you know everything I know. What is important is you have what you require to be a great shooter and scorer.

To Know Is to Do

Now it's up to you to put it in practice.

Some readers may fail to understand the value of the information presented here. But others will get it. The critical concept is easily understood: Controlling the centerline of the ball relative to the center of the basket at the moment of release will result in the ball going straight. There are more ways to control the ball's centerline than have been presented here. Feel free to experiment to find what works best for you.

The more you think about this material, the more sense it will make to you. The more familiar you become with these concepts, the more you will find this stuff works.

This approach is a departure from traditional shooting instruction, but it will open a new universe of opportunities for you, especially when you come to understand these concepts at a deep level. You will be able to make shots in a wide variety of ways.

KPT

Keep in mind all it takes to become great at anything is knowledge, practice and time.

By reading this book, you have improved your knowledge, which will lead to an improvement in your shooting. An increase in knowledge will give you better mental pictures or representations.[1] Better mental representations lead to better practice. And better practice leads to an improvement in your shooting—and still better mental representations. It is a continuous cycle.

Becoming a great shooter is a process. Learn to enjoy the process of getting better. It is not a matter of doing your best one time. It is more a matter of finding out how good your best can be.

Yes, You Can

I vividly recall one time, after completing an arduous task as a boy, my mother looked me in the eye and said, "Bobby, you really can do anything when you put your mind to it."

She was right.

And you can too.

Notes:

Intro:

1. *Figuring 'The Physics of Basketball'.* NPR, NPR, 30 Mar. 2007, www.npr.org/templates/story/story.php?storyId=9243416.

Chapter 1: The Problem With Mechanics

1. Beard, Rod. *Andre Drummond 'open' to going underhanded at the line. Detroit News,* The Detroit News, 28 Apr. 2016, www.detroitnews.com/story/sports/nba/pistons/2016/04/28/drummond-could-go-going-underhanded-line/83655618/.

2. Farrell, Perry A. *Pistons hire Dave Hopla as shooting coach. Detroit Free Press,* Detroit Free Press, 26 June 2015, www.freep.com/story/sports/nba/pistons/2015/06/26/detroit-pistons-dave-hopla/29348935/.

3. Dwight Howard: *Orlando Magic all-Star Dwight Howard hires Ed Palubinskas to be his shooting coach. Tribunedigital-Orlandosentinel,* 25 Aug. 2011, articles.orlandosentinel.com/2011-08-25/sports/os-dwight-howard-free-throws-0826-20110825_1_dwight-howard-ed-palubinskas-orlando-magic.

4. *The Last Season: A Team in Search of Its Soul. Wikipedia,* Wikimedia Foundation, en.wikipedia.org/wiki/The_Last_Season:_A_Team_in_Search_of_Its_Soul.

5. *Dwight Howard Stats. Basketball-Reference.com,* www.basketball-reference.com/players/h/howardw01.html.

6. *Shot Caller. SLAMonline,* 28 Mar. 2016, www.slamonline.com/the-magazine/features/ted-st-martin-free-throw-record/#7K-W7yvDJbJJFkYce.97.

7. Hopla, Dave. *Basketball shooting.* Champaign, IL, Human Kinetics, 2012.

8. *Palubinskas-Freethrow ideal shot pocket. YouTube,* YouTube, 4 Feb. 2011, www.youtube.com/watch?v=FHXN-eMuz5g.

9. Barnhardt, Wilton. *HE THROWS FREE THROWS*

BY THE SCORE. SI.com, 13 Oct. 2015, www.si.com/vault/1991/01/28/123518/he-throws-free-throws-by-the-score-floridian-ted-st-martin-has-parlayed-an-astonishing-talent-into-a-unique-career.

10. *Attention and Motor Skill Learning* - Gabriele Wulf. *Human-Kinetics*, www.humankinetics.com/products/all-products/attention-and-motor-skill-learning

11. LearnTheRealSwing. *Out with the old paradigm, in with the new... Internal Focus vs External Focus in ball-Striking. YouTube*, YouTube, 27 Sept. 2014, www.youtube.com/watch?v=ey9kY6bWCh4.

12. Dweck, Carol. *Mindset: The New Psychology of Success*. New York: Ballantine Books, 2006. Print.

Chapter 2: Uncharted Waters

1. Sharman, Bill. *Sharman on basketball shooting, by Bill Sharman*. Englewood Cliffs, NJ, Prentice-Hall, 1967.

2. Fontanella, John Joseph. *The Physics of Basketball*. Baltimore: Johns Hopkins UP, 2006. Print.

3. Coyle, Daniel. *The Talent Code: greatness isn't born. It's grown*. London, Arrow, 2010.

4. Ericsson, K. Anders. *Peak: secrets from the new science of expertise*. Boston, Houghton Mifflin Harcourt, 2016.

5. Vickers, Joan N. *Perception, cognition, and decision training: the quiet eye in action*. Champaign, IL, Human Kinetics, 2007.

6. IJSEP, 2005, 3, 197-221 Education of Attention in www.bing.com/cr?IG=5A68D28BC38A45CEABD0DB5811D6AE7C&CID=23E09C9898C1605C30D7966699C761FA&rd=1&h=5HmxWvoB-z3c_BKX5CgnjmpJYgYk1ehjVVcTFQ16GYU&v=1&r=http%3a%2f%2fwww.basketball.

7. *The University of Pittsburgh - MSRC*, www.pitt.edu/~msrc/publications/2002/li_69.html.

8. Wolfe, Scott W., et al. *The Dart-Throwing Motion of the Wrist: Is*

It Unique to Humans? The Journal of Hand Surgery, U.S. National Library of Medicine, Nov. 2006, www.ncbi.nlm.nih.gov/pmc/articles/PMC3260558/.

9. Gawande, Atul. *The Coach in the Operating Room. The New Yorker*, The New Yorker, 19 June 2017, www.newyorker.com/magazine/2011/10/03/personal-best.

10. Gladwell, Malcolm. *David and Goliath: underdogs, misfits, and the art of battling giants*. New York, Back Bay Books / Little, Brown and Company, 2015.

11. Chang, Chein-Wei, et al. *Increased Carrying Angle is a Risk Factor for Nontraumatic Ulnar Neuropathy at the Elbow. SpringerLink*, Springer-Verlag, 28 May 2008, link.springer.com/article/10.1007/s11999-008-0308-2.

12. Waitzkin, Josh. *The art of learning: an inner journey to optimal performance*. New York, Free Press, 2008.

13. gibbyworldpolice. *Winning Basketball Bird. YouTube*, YouTube, 25 Aug. 2009, www.youtube.com/watch?v=tporD7Cef7Q.

14. *Figuring 'The Physics of Basketball'. NPR*, NPR, 30 Mar. 2007, www.npr.org/templates/story/story.php?storyId=9243416.

15. Goldaper, Sam. *Hank Luisetti, 86, Innovator Of Basketball's One-Hander.* The New York Times, 22 Dec. 2002, www.nytimes.com/2002/12/23/sports/hank-luisetti-86-innovator-of-basketball-s-one-hander.html.

16. Lewis, Michael. *Moneyball: the art of winning an unfair game*. New York, W.W. Norton, 2013.

Chapter 3: The Secret of the Release

1. Oliver, Dean. *Basketball on Paper Rules and Tools for Performance Analysis*. Washington, D.C.: Brassey's, 2004. Print

2. Chase, Chris. *Nine Amazing Stats About San Antonio's Dominant 2014 NBA Championship. USA Today* 16 June 2014. Web. 1 May 2015.

3. Graves, Will. *Pittsburgh Drills No. 12 North Carolina 89-76*. CBS Sports, 14 Feb. 2015. Web. 6 May 2015. <http://www.cbssports.com/collegebasketball/gametracker/recap/NCAAB_20150214_UNC@PITT>

4. Haruno, Masahiko, and Daniel M. Wolpert. *Optimal Control of Redundant Muscles in Step-Tracking Wrist Movements*. Journal of Neurophysiology, American Physiological Society, 1 Dec. 2005, jn.physiology.org/content/94/6/4244.

5. Poteet, James L. *The Paradox of the Free Throw*. 1999. Print.

6. *Increased Carrying Angle Is a Risk Factor for Nontraumatic Ulnar Neuropathy at the Elbow Result Filters*. National Center for Biotechnology Information. U.S. National Library of Medicine. Web. 12 July 2015.

7. Nathan, Alec. *Frank Mason III Wins 2016-17 Wooden Award. Bleacher Report*, Bleacher Report, 12 Apr. 2017, bleacherreport.com/articles/2701831-frank-mason-iii-wins-2016-17-wooden-award.

8. Fontanella, John Joseph. *The Physics of Basketball*. Baltimore: Johns Hopkins UP, 2006. Print.

9. Wulf, Gabriele. *Attention and Motor Skill Learning*. Champaign, IL: Human Kinetics, 2007. Print.

10. Brown, Peter C. *Make It Stick: The Science of Successful Learning*. Print

11. Coyle, Daniel. *The Talent Code: Greatness Isn't Born, Its Grown. Here's How*. London, Arrow, 2010.

12. Miller, S. and Bartlett, R.M. (1996). *The Relationship between Basketball Shooting Kinematics, Distance and Playing Position*. Journal of Sports Sciences, 14, 243-253

Chapter 4: Objectives

1. Brancazio, Peter J. *Physics of basketball. American Journal of Physics*, aapt.scitation.org/doi/10.1119/1.12511.

2. Haruno, Masahiko, and Daniel M. Wolpert. *Optimal Control of Redundant Muscles in Step-Tracking Wrist Movements. Journal of Neurophysiology*, American Physiological Society, 1 Dec. 2005, jn.physiology.org/content/94/6/4244.

3. K, Yuvaraj Babu., and P. Saraswathi. *A Study on influence of Wrist Joint Position on Grip strength in normal Adult Male Individuals. International Journal of Drug Development and Research*, IMedPub, 8 May 2015, www.ijddr.in/drug-development/a-study-on-influence-of-wrist-joint-position-on-grip-strength-in-normaladult-male-individuals.php?aid=5637.

Chapter 5: Be a Straight-shooter

1. Branch, John. *For Free Throws, 50 Years of Practice Is No Help.* The New York Times, 3 Mar. 2009, www.nytimes.com/2009/03/04/sports/basketball/04freethrow.html.

2. jsnlee102. *Volkswagen Commercial - Father Son Baseball/Catch. YouTube*, YouTube, 8 Jan. 2013, www.youtube.com/watch?v=VxAo8_JySkM.

3. Sharman, Bill. *Sharman on basketball shooting.* Englewood Cliffs, NJ, Prentice-Hall, 1965.

4. Alfonso, Tony, et al. "The Art of Shooting." Hoops U. Basketball, 29 June 2011, hoopsu.com/art-of-shooting-lehmann/.

5. proshotcoach. "*Basketball Shooting Components: FINGER.*" *YouTube*, YouTube, 23 June 2016, www.youtube.com/watch?v=wdax27v2I2I %28the Finger %E2%80%93 Pro Shot%29.

6. Chang, Chein-Wei, et al. "*Increased Carrying Angle is a Risk Factor for Nontraumatic Ulnar Neuropathy at the Elbow.*" *SpringerLink*, Springer-Verlag, 28 May 2008, link.springer.com/article/10.1007/s11999-008-0308-2.

7. gibbyworldpolice. *Winning Basketball Bird. YouTube*, YouTube, 25 Aug. 2009, www.youtube.com/watch?v=tporD7Cef7Q.

8. Bartlett, R, et al. *Is Movement Variability Important for Sports Biomechanists? Sports biomechanics.*, U.S. National Library of Medicine, May 2007, www.ncbi.nlm.nih.gov/pubmed/17892098.

Chapter 8: Long and Short

1. Fontanella, John Joseph. *The Physics of Basketball*. Baltimore: Johns Hopkins UP, 2006. Print.

2. Bhatia, Aatish. *Galileo Got Game: 5 Things You Didn't Know About the Physics of Basketball*. *Wired*, Conde Nast, 3 June 2017, www.wired.com/2014/04/basketball-physics/.

3. Barzykina, Irina. *The Physics of an Optimal Basketball Free Throw*. *[1702.07234] The physics of an optimal basketball free throw*, 21 Feb. 2017, arxiv.org/abs/1702.07234.

4. O'Connor, Kevin. *A Flick of the Wrist*. *The Ringer*, The Ringer, 10 Oct. 2016, www.theringer.com/nba/2016/10/10/16077030/nba-shooting-coaches-kent-bazemore-kawhi-leonard-8660e9939680.

5. Lewis, Michael. *Moneyball: the art of winning an unfair game*. New York, W.W. Norton, 2013.

6. *In basketball, shooting angle has a big effect on the chances of scoring*. *The Washington Post*, WP Company, 16 Mar. 2010, www.washingtonpost.com/wp-dyn/content/article/2010/03/15/AR2010031502017.html.

7. Brancazio, Peter J., et al. *Sport Science*. By Peter J. Brancazio, www.goodreads.com/book/show/1699681.Sport_Science.

8. Venkadesan, M., and L. Mahadevan. *Optimal strategies for throwing accurately*. *Royal Society Open Science*, The Royal Society Publishing, Apr. 2017, www.ncbi.nlm.nih.gov/pmc/articles/PMC5414278/.

Chapter 9: Ways to Miss

1. Fontanella, John Joseph. *The Physics of Basketball*. Baltimore: Johns Hopkins UP, 2006. Print.

Chapter 11: Sidespin

1. Fontanella, John Joseph. *The Physics of Basketball*. Baltimore: Johns Hopkins UP, 2006. Print.

Chapter 12: Bank

1. *Borgata Free-Throw Tournament A Hit | Casino Gambling News. CalvinAyre.com*, 26 Oct. 2016, calvinayre.com/2015/03/22/casino/borgata-free-throw-tournament-a-hit/.

2. Silverberg, Larry M, et al. *Optimal Targets for the Bank Shot in Men's Basketball. Journal of Quantitative Analysis in Sports*, De Gruyter, 2 Apr. 2013, www.degruyter.com/view/j/jqas.2011.7.1/jqas.2011.7.1.1299/jqas.2011.7.1.1299.xml.

   Credit Larry Silverberg, Chau Tran, and Taylor Adams for aim diagrams

Chapter 13: Free Throws

1. Hall, Brandon. *Elena Delle Donne's 7 Steps to a Perfect Free Throw. STACK*, 28 Oct. 2015, www.stack.com/a/elena-delle-donnes-7-steps-to-a-perfect-free-throw.

2. *New Insight into the Neuroscience of Choking under Pressure - 11/04/2014. Johns Hopkins Medicine, based in Baltimore, Maryland*, www.hopkinsmedicine.org/news/media/releases/new_insight_into_the_neuroscience_of_choking_under_pressure.

3. Cuda, Gretchen. *Just Breathe: Body Has A Built-In Stress Reliever. NPR*, NPR, 6 Dec. 2010, www.npr.org/2010/12/06/131734718/just-breathe-body-has-a-built-in-stress-reliever.

4. *Why Deep Breathing Is the Fastest Way to Calm You Down. Time*, Time, time.com/4718723/deep-breathing-meditation-calm-anxiety/.

5. Khazan, Olga. *Male and Female Brains Really Are Built Differently. The Atlantic*, Atlantic Media Company, 2 Dec. 2013, www.theatlantic.com/health/archive/2013/12/male-and-female-brains-really-are-built-differently/281962/.

6. O'Donnell, Erin. *A Better Path to High Performance. Harvard Magazine*, 24 Aug. 2016, harvardmagazine.com/2014/05/a-better-path-to-high-performance.

7. Pillay, Srini. *How to Deal with Unfamiliar Situations. Harvard Business Review*, 1 Nov. 2014, hbr.org/2014/03/how-to-deal-with-unfamiliar-situations.

Chapter 15: Downtown

1. Gladwell, Malcolm. *What the dog saw and other adventures.* London, Penguin Books, 2010.

2. Digit Ratio: *A Pointer to Fertility, Behavior, and Health (A volume in the Rutgers Series in Human Evolution*, edited by Robert Trivers.) Paperback – February 1, 2002 by John Manning (Author)

3. Bible, Adam. *How golfer Tim Burke became a long-Drive champion. Men's Fitness*, Men's Fitness, 30 Aug. 2017, www.mensfitness.com/sports/golf/how-golfer-tim-burke-became-long-drive-champion.

Chapter 16: Dr. Tom

1. Amberry, Tom, and Philip Reed. *Free throw: 7 steps to success at the free throw line.* New York, Harper Perennial, 1996.

Chapter 17: Tall vs Short

1. Ballard, Chris. *The art of a beautiful game: the thinking fans tour of the NBA.* New York, Simon & Schuster, 2009.

2. Fontanella, John Joseph. *The Physics of Basketball.* Baltimore: Johns Hopkins UP, 2006. Print.

Chapter 18: High or Low

1. Brancazio, Peter J., et al. *Sport Science.* By Peter J. Brancazio, www.goodreads.com/book/show/1699681.Sport_Science

Chapter 19: Exercises

1. Ericsson, K. Anders. *Peak: secrets from the new science of expertise.* Boston, Houghton Mifflin Harcourt, 2016.

2. *The 4 Rituals That Will Make You An Expert At Anything. Barking Up The Wrong Tree*, 3 May 2016, www.bakadesuyo.com/2016/03/expert/.

Chapter 20: Closing Comments
1. Ericsson, K. Anders. *Peak: secrets from the new science of expertise.* Boston, Houghton Mifflin Harcourt, 2016.

Acknowledgments

This book was set in motion by *The Physics of Basketball* by John Fontanella; which altered my course and resulted in this work. Without it, I would still be grasping at straws. John's help and friendship managed to keep me on the right track, and for that, I am enormously grateful.

A sincere thank you to all the friends, players, coaches and family who have supported me throughout this journey. Heartfelt thanks to those who took time to help assist in record-setting attempts. Also, I am indebted to the sports writers and media who took an interest in my story. John Marshall, Brad Fanning, Steve Ridgeway, John Branch, Kevin Haskins, and Julie Perry deserve special mention.

Thanks to all the players I had the privilege of coaching over the years. Although it is not how the world works, I wish I had known this stuff then.

Special thanks to Phil Reed, who read my core chapter and said, "I think there is a book here, and you are just the person to write it." That comment provided all the inspiration I needed. Also, Phil read and edited the manuscript and helped teach a novice (as best he could with what he had to work with) the basics.

I am indebted to Professor Larry Silverberg for supplying the chapter covering the physics of the free throw. Professor Silverberg read the manuscript and wrote a review that was so thorough that it prompted the idea for a special chapter—which he was kind enough to supply. His contribution from an academic perspective is invaluable.

Credit and thanks to Garrett Steinlage and Justin Smith for providing the diagrams in this manuscript.

I am grateful to Jolene Fairchild, Sarah Bostic, and Allan Wegner for reading this manuscript and providing editing comments.

Blessings to Kathy Lorkovic and Marilyn Iturri for their editing prowess. They took a rough draft and made it readable.

I appreciate Jolene Brown directing us to Steve Himes and Telemachus Press. Working with Steve and MaryAnn Nocco was a pleasure and enabled this work to come to fruition.

My deepest thanks to Nikki, Jennifer, Danny, and Sarah along with our grandkids: Brynlee, Emily, Lucas, and Levi. I love you.

Last and most important, thank you to my wife Connie for believing in me. Throughout the long and difficult process of figuring this stuff out, the duties that I shirked at home fell squarely on her. Connie has been my sounding board and partner in this journey. Her encouragement, assistance, love, and support made this book possible.

About the Author

Bob Fisher is an internationally recognized basketball shooting expert. He produced the video, *Secrets of Shooting* (in 2008) and currently holds 14 Guinness World Records™ in basketball shooting. Bob has appeared on *The Tonight Show with Jay Leno*, *Inside Edition*, and was a feature story on the front page of the *New York Times*. Bob is the CEO of Fisher Sharp Shooters, LLC and gives presentations and teaches shooting to rapidly-improving players around the country

CPSIA information can be obtained
at www.ICGtesting.com
Printed in the USA
BVHW021526200820
586925BV00020B/2494